EAI/Springer Innovations in Communication and Computing

Series editor

Imrich Chlamtac, European Alliance for Innovation, Ghent, Belgium

Editor's Note

The impact of information technologies is creating a new world yet not fully understood. The extent and speed of economic, life style and social changes already perceived in everyday life is hard to estimate without understanding the technological driving forces behind it. This series presents contributed volumes featuring the latest research and development in the various information engineering technologies that play a key role in this process.

The range of topics, focusing primarily on communications and computing engineering include, but are not limited to, wireless networks; mobile communication; design and learning; gaming; interaction; e-health and pervasive healthcare; energy management; smart grids; internet of things; cognitive radio networks; computation; cloud computing; ubiquitous connectivity, and in mode general smart living, smart cities, Internet of Things and more. The series publishes a combination of expanded papers selected from hosted and sponsored European Alliance for Innovation (EAI) conferences that present cutting edge, global research as well as provide new perspectives on traditional related engineering fields. This content, complemented with open calls for contribution of book titles and individual chapters, together maintain Springer's and EAI's high standards of academic excellence. The audience for the books consists of researchers, industry professionals, advanced level students as well as practitioners in related fields of activity include information and communication specialists, security experts, economists, urban planners, doctors, and in general representatives in all those walks of life affected ad contributing to the information revolution.

Indexing: This series is indexed in Scopus, Ei Compendex, and zbMATH.

About EAI

EAI is a grassroots member organization initiated through cooperation between businesses, public, private and government organizations to address the global challenges of Europe's future competitiveness and link the European Research community with its counterparts around the globe. EAI reaches out to hundreds of thousands of individual subscribers on all continents and collaborates with an institutional member base including Fortune 500 companies, government organizations, and educational institutions, provide a free research and innovation platform.

Through its open free membership model EAI promotes a new research and innovation culture based on collaboration, connectivity and recognition of excellence by community.

More information about this series at http://www.springer.com/series/15427

Prashant Johri • Mario José Diván
Ruqaiya Khanam • Marcelo Marciszack
Adrián Will

Editors

Trends and Advancements of Image Processing and its Applications

Editors
Prashant Johri
Plot No 2, Sector 17 A, Yamuna Expressway
School of Computing Science &
Engineering, Galgotias University
G.B.Nagar, Greater Noida
Uttar Pradesh, India

Ruqaiya Khanam
Computer Science and Engineering
Department, SET
Sharda University
Greater Noida, Uttar Pradesh, India

Adrián Will
Information System Department
National Technological University -
Tucumán Regional Faculty
Tucumán, Tucumán, Argentina

Mario José Diván
Data Science Research Group at
Economy School
National University of La Pampa
Santa Rosa, Argentina

Marcelo Marciszack
Information System Department, Center for
Research, Development, and Transfer of
Information Systems (CIDS)
National Technological University -
Córdoba Regional Faculty
Córdoba, Argentina

ISSN 2522-8595   ISSN 2522-8609   (electronic)
EAI/Springer Innovations in Communication and Computing
ISBN 978-3-030-75947-6   ISBN 978-3-030-75945-2   (eBook)
https://doi.org/10.1007/978-3-030-75945-2

This Springer imprint is published by the registered company Springer Nature Switzerland AG
The registered company address is: Gewerbestrasse 11, 6330 Cham, Switzerland

To our parents:
Sh. A. N. Johri and Mithlesh Johri
Anita Koller
Bilqis Hamid
Gladys Beatriz Bussolini and Enrique Otto
Marciszack

and to our families:
Methily Johri, Aniket Johri, and
Mamta Saxena
José Ignacio, Santiago Agustín, and
María Laura
Kashif A. Ahmad, Maaz, Almaas, and Fawaz
Erika, Federica, and María Alejandra
Amelia, Iván, and Jeremías

Preface

Image processing is a great tool for many applications in our day-to-day life. Anyone could be captivated by the variety of forms, colors, textures, and motion among other perceptible aspects in the world. Human beings have a great ability to acquire, integrate, and interpret the information surrounding them. It is a challenging task to impart these capabilities to a machine in a way to interpret the visual information deeply embedded in a still image, moving image, video, and graphics in the sensory world. As a result, there is a lot of application areas of digital image processing, such as multimedia computing, biomedical imaging, biometrics, security image data communication, pattern recognition, texture understanding, remote sensing, and image compression content-based image retrieval.

This book is intended to cover current technological innovations and applications in the emerging field of image processing and analysis techniques with application in the clinical industry, remote sensing, forensics, astronomy, manufacturing, defense, and many more that depend on image storage, display, and information disclosure about the world around us. The book presents the concepts and techniques of remote sensing, such as image mining and geographical and agricultural resources, medical image detection and diagnoses, image recognition and analysis, artificial intelligence, computer vision, machine learning, deep learning-based image analysis, pattern recognition and capsule networks in object tracking, remote sensing, moving object tracking, and wavelet transformation.

The book targets undergraduate, graduate, and post-graduate students, researchers, academicians, policymakers, government officials, academicians, technocrats, industry research professionals who are currently working in the field of academic research, and the research industry to improve the lifespan of the general public.

The book is organized into 16 chapters. Chapter 1 describes the application of convolutional neural network (CNN) for classifying COVID-19 based on computerized tomography scans. Chapter 2 outlines challenges related to image processing on mobile devices. Chapter 3 synthesizes a proposal applied to Smart City, using Internet of Things (IoT) and image processing. Chapter 4 proposes an application of artificial intelligence (AI) for dental image analysis. Chapter 5 analyzes the feature extraction techniques towards image processing of the plant extracts. Chapter 6 proposes a median filter based on entropy to remove noise in images. Chapters 7 and 8 describe the application of deep learning models for predicting COVID-19 and chronic myeloid leukemia, respectively. Chapter 9 addresses an automatic bean classification system based on visual features. Chapter 10 describes a supervised model to classify human sperm head. Chapter 11 proposes the future contribution of computational vision in the detection of maturity states of medicinal cannabis in Colombia. Chapter 12 describes the detection of brain tumor region based on magnetic resonance images (MRI) using clustering algorithms. Chapter 13 proposes the use of CNN for estimating human posture through pictures. Chapter 14 addresses a human skin detection technique supported by color models. Chapter 15 proposes a study of improved methods on image inpainting.

We have tried to gather a broad field of current and representative applications related to different perspectives of image processing. In such a sense, we want to express our gratitude to all of our contributors and friends, who brought their opinion and researches with different viewpoints, enriching the subject treatment.

Greater Noida, Uttar Pradesh, India Prashant Johri
Santa Rosa, Argentina Mario José Diván
Greater Noida, Uttar Pradesh, India Ruqaiya Khanam
Córdoba, Argentina Marcelo Marciszack
Tucumán, Argentina Adrián Will

Acknowledgments

Our special thanks to God, our families, and friends for supporting, enlightening, and helping us every day.

This work has been partially supported by the CD 312/18 project at the National University of La Pampa (Argentina) and Grant UTN CCUTITU0007896TC from the National Technological University of Argentina.

Our gratitude to the Center for Research, Development, and Transfer of Information Systems (CIDS) of the Córdoba Regional Faculty of the National Technological University (Argentina).

Contents

Part I
Recent Trends and Advancements of Image Processing and its Applications

Using Convolutional Neural Networks for Classifying COVID-19 in Computerized Tomography Scans

Lúcio Flávio de Jesus Silva, Elilson dos Santos, and Omar Andres Carmona Cortes

1 Introduction

Computer-aided diagnosis (CAD) systems, according to Esteves et al. [1], "generally involve a classification step, which determines, for example, the presence or absence of a disease of interest." Thus, early detection of a disease can provide sufficient time for successful treatment or action.

CAD systems fit into a field called Medicine 4.0 [2]. One of the applications includes the combination of innovative artificial intelligence technologies, including machine learning algorithms, to develop support for clinical decisions. These decision support systems are related to procedures to improve medical decisions, providing evidence-based information at the time of contact between doctor and patient or even when deciding on treatment [3].

Machine learning (ML), according to Monard and Baranauskas [4], "is an Artificial Intelligence research field that studies the development of methods capable of extracting concepts (knowledge) from data samples." In general, ML algorithms are used in order to generate classifiers for a set of examples. Classification means the process of assigning, to a given data, the label of one (or more) class to which it belongs. In this sense, ML techniques are used to induce a classifier that, based on examples, is able to predict the class of new data of the amount in which it was trained.

L. F. de Jesus Silva · E. dos Santos
Universidade Estadual do Maranhão (UEMA), Programa de Pós-Graduação em Engenharia de Computação e Sistemas (PECS), Av. Cidade Universitária Paulo VI, Av. Lourenço Vieira da Silva, São Luís, MA, Brazil

O. A. C. Cortes (✉)
Instituto Federal de Educação, Ciência e Tecnologia do Maranhão (IFMA), Computer Department (DComp), São Luis, MA, Brazil
e-mail: omar@ifma.edu.br

© Springer Nature Switzerland AG 2022
P. Johri et al. (eds.), *Trends and Advancements of Image Processing and its Applications*, EAI/Springer Innovations in Communication and Computing,
https://doi.org/10.1007/978-3-030-75945-2_1

Machine learning approaches have evolved into deep learning approaches, which are a more robust and efficient way of dealing with the huge amounts of data generated from modern discovery approaches [5]. However, deep neural networks, especially convolutional neural networks (CNNs) that work with images, have many layers and connections, making them computationally complex for the complete training of the network. However, a solution called transfer learning can be applied to help with this problem. The idea is to use prior knowledge to solve similar problems, as humans do. Thus, transfer learning is a method of reusing a model or knowledge for another related task [6].

1.1   Problem Definition

The outbreak of coronavirus 2 2019 (COVID-19) of severe acute respiratory syndrome (SARS-CoV-2) has led to millions of people being infected and thousands of deaths worldwide. The World Health Organization (WHO) released the Situation Report reporting 41,570,883 cases and 1,134,940 deaths worldwide by October 23, 2020 [7]. The outbreak caused the health systems to collapse and questioned society's preparedness in the face of unknown virus pandemics, where immediate diagnosis and isolation could prevent a mass spread.

With the development of Computer-Aided Disease Diagnostic Systems, the possibility arose to determine, for example, the presence or absence of a disease of interest. Deep learning algorithms, mainly convolutional neural networks (CNNs), have been used successfully in the processing and analysis of digital images, including medical images [8]. The study on the performance of several CNN architectures in the analysis and classification of medical images can assist professionals and researchers in the development of these systems, applying the architecture that achieved the best performance in solving a certain problem.

In this work, 11 deep learning architectures will be presented in the task of computational analysis of medical images in order to compare the classifiers in the diagnostic task of COVID-19 analyzing computed tomography images. The architectures used are DenseNet121, DenseNet169, DenseNet201, Resnet50, Resnet50 v2, VGG16, VGG19, MobileNet, MobileNetV2, Xception, and Inception v3. To carry out the experiments, a database composed of 2477 computed tomography images was used, which is used to detect lung diseases, 1250 of which tested positive for COVID-19 and 1227 had a negative diagnosis.

1.2   Formulation of Hypotheses

In view of technological advances in the field of medicine, the deep learning algorithms have gradually helped professionals in the area in the analysis of clinical images; however, there is a lack of studies that identify which architectures are most suitable for analyzing these images. Therefore, which deep learning architectures performed better in the computational analysis of medical images?

1.3 Objective

Analyze and compare the performance of different deep learning architectures using transfer learning in the diagnosis of different pathologies in different databases, initially from COVID-19.

Specific Objective

The specific objectives of this work are:

- Carry out a bibliographic survey on different architectures of deep learning and transfer learning.
- Implement means for the pre-processing of computer tomography (CT) data.
- Determine techniques that assist in training the models.
- Compare the performance of the different deep learning architectures in the diagnosis of pathologies.

1.4 Methodology

To carry out this work, it was divided into five major activities: bibliographic survey, collection and analysis of sources, specification of requirements for the architectures used, architectural training, and performance analysis between different architectures.

The search for this content was carried out in books by renowned authors and cited in the context of research related to machine learning and deep learning; in addition to scientific articles with publications in magazines and university sites, search engines such as Google Scholar were also used. Moreover, specific words were used as a way to specify the search for this research material, such as machine learning, deep learning, artificial neural networks, image classification, and tomographic images. These descriptors generated significant assistance in the development of scientific research.

The main difficulty encountered was obtaining a robust dataset containing tomographic images of patients diagnosed with a specific pathology. However, there is a dataset called the SARS-COV-2 Ct-Scan Dataset [9], which united researchers with the study data, and this dataset was used for the first stage of this project.

A team of researchers from Qatar University, Doha, Qatar, and the University of Dhaka, Bangladesh, together with their collaborators from Pakistan and Malaysia in collaboration with doctors, have also created a database of chest X-ray images called COVID -19 Radiography Database containing positive cases of COVID-19 together with standard cases and images of viral pneumonia [10].

2 Literature Review

2.1 Machine Learning

According to Mitchell [11] "Machine Learning research seeks to develop techniques capable of extracting knowledge from examples of data." Machine learning algorithms are identified, the purpose of which is to enable a machine to be able to interpret new information and classify it appropriately after a certain training with a certain cluster of data by which the instances have family classification, from a generalization of what was shown previously. According to Libralão et al. [12], there are three paradigms that suggest how to learn the machine learning algorithm:

- Supervised: needs supervisors to achieve the best model intended in the training phase.
- Unsupervised: it allows the AM algorithm to learn to group the inputs taking into account a measure of similarity between them.
- For reinforcement: learning happens through bonuses, according to the performance achieved.

Also, according to Libralão et al. [12] most of the ML algorithms are inspired by biological systems (artificial neural networks (ANNs) and genetic algorithms), symbolic learning (decision trees), cognitive processes (case-based reasoning), and theory statistics (SVMs).

Artificial Neural Networks

ANN is an area of research that works with the simulation of human cognitive skills. Machines are developed that are capable of intelligent behavior, as if they were human behaviors. Human intelligence is the most developed among other creatures and the region responsible for this intelligence in humans is the brain [13].

Still according to Rauber [13], the essential elements are neurons, interconnected in networks, which allows the sharing of information between them, forming biological intelligence. An apparent interest that arises from these facts is the effort to copy the way the brain works in a computational environment. This means that the research tries to understand the functioning of the intelligence inhibited in neurons and diagram it in an artificial structure, such as a junction of hardware and software, thus converting biological neural networks into artificial neural networks.

Neuroscience Inspiration

What are the particularities of the human brain that enable it to behave intelligently? The following topics reflect the most significant characteristics that are mainly attractive to be used in an artificial neural network, according to Rauber [13]:

- Robustness and fault tolerance: The exclusion of neurons does not affect general functionality.
- Learning capacity: The brain has the ability to learn new tasks that have never been done before.
- Uncertain information processing: Even if the knowledge provided is incomplete, affected by noise, or relatively contradictory, precise reasoning is still possible.
- Parallelism: A huge number of neurons are active simultaneously. There is no restriction on a processor that must work with information after another.

Figure 1 shows the summarized model of a real neuron. The neuron is a cell that has a nucleus and a body where chemical and electrical behaviors demonstrate the treatment of information. The sum information is outputted by electrical impulses that diffuse through the axon. At the end of the axon are numerous branches that share information for neighboring neurons. The junction with other neurons is done through synapses that are linked to a dendrite of the receptor neuron. The synapse sends a chemical when it is alerted by the axon pulse. The substance spreads between synapse and dendrite, making the coupling between two neighboring neurons. According to the alerts (or inhibitions) that neighboring cells send to the cell under consideration, it treats the information once again and transmits it through its axon [13].

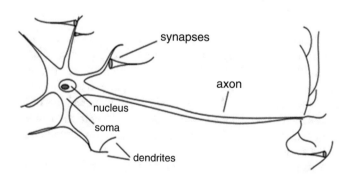

Fig. 1 Biological neuron, Rauber [13]

Brief History of Neurocomputing

An important point in the history of ANNs was the demonstration of a model of an artificial neuron by McCulloch and Pitts [14]. The works in this line of research ended in perceptron's conception by Rosenblatt [15] and, in a similar model, adaline by Widrow and Hoff [16]. Perceptron has the ability to classify between classes that are linearly separable. It was used to recognize characters, for example. This application was executed on a machine called MARK I PERCEPTRON, and it caused a great euphoria certainly exorbitant in relation to the imagination of the abilities of future intelligent robots.

Fundamentals

According to Prampero [17] "Artificial Neural Networks (ANNs) apply a mathematical model influenced by the neural architecture of living beings, contracting knowledge through experience."

Tatibana and Kaetsu [18] declare that "Artificial Neural Networks are formed by a grouping of nodes, which imitate the role of neurons, linked by a propagation principle. Each node acquires its inputs with the related weights, coming from other nodes, or from an external interaction. The input layer has a specific node, called a bias, which helps to increase the degrees of freedom, enabling a better adaptation of the network."

Also, on these inputs, an activation function is used, which uses a weighted sum as an argument in the network inputs. "The activation point of a node is defined by the activation function, usually a sigmoidal function or a step function" [19]. A layer that acquires signals from the outside environment is called an input layer and a layer that expresses signals to the environment is called an output layer. All others are defined as hidden layers and do not connect directly with the environment [20].

Learning Paradigms

According to Rauber [13], once the neural network has been determined, it must be trained. This means that the degrees of autonomy that the network has to solve the task under consideration must be adapted in the best way. Usually, this means that weights must be changed between neuron i and neuron j, according to an algorithm. A finite set T of n training samples is available to adapt the weights during the training of the network. A fundamental distinction in relation to the learning paradigm that is relevant for all types of systems with adaptation techniques is supervised learning and unsupervised learning.

In supervised learning, each training sample is accompanied by a cost that is the desired cost. This means that training set T is composed of n pairs of samples (xp yp) where $T = \{(xp\ yp)\}\ np = 1$. The dimension of the input vector is D, i.e., the input variables are associated with one multidimensional value, usually part of the

Fig. 2 Linear regression, Rauber [13]

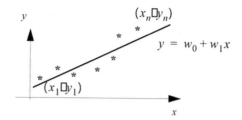

real numbers: x = (x1… xj… xD)T, xj R. (The transposition (.)T is used to save space by registering a vector on a line.) The output variables are gathered in an output vector y = (y1… yi… yc) T [13].

According to Rauber [13], a demonstration of a supervised learning task is linear regression. The one-dimensional case is used to simplify the illustration, as shown in Fig. 2. In this problem the training set consists of pairs of real numbers (xp yp). The learning objective is the delimitation of coefficients w0 and w1 of the line y = w0 + w1x. The learning algorithm promotes to minimize the discrepancy between the desired value yp and the value that is the answer y' = w0 + w1xp of the system, and this on average for each sample (xp yp).

In unsupervised learning, when there is only one information available, and that information is the values (xp) the learning task is to find correlations between the training examples T = {(xp)} np = 1. The number of groups or classes is not defined a priori. This means that the network has to find considerable statistical attributes, and it has to develop its own interpretation of the stimuli that enter the network. A synonym for unsupervised learning is clustering [13].

An example from the medical field is the detection of diseases using images, such as X-ray images. There are places within the image that can be attributed to the same material, such as bone. The number of materials (from agglomerations) is initially unknown. The objective of the system is to find the number of different materials and at the same time group each point of the image for the respective material. The entrance to the network could be the points of the image, for example, a small interval of 5 by 5 points. The expected response from the network would be the material to which this region of the image belongs [13].

2.2 Deep Learning

Deep learning allows computational models composed of several layers of processing to learn representations of data with varying levels of abstraction. According to Lecun [21] these methods have drastically improved the state of the art in speech recognition, object recognition, object detection, and many other domains. Deep learning discovers the intricate structure in large datasets using the backpropagation algorithm to indicate how a machine should change its internal parameters, which are used to calculate the representation in each layer from the representation in the previous layer. Deep convolutional networks brought advances in image, video, voice, and audio processing, while recurring networks illuminated sequential data, such as text and speech.

Convolutional Neural Networks

Convolutional neural networks are a variation of multilayer perceptron (MLP). By biological inspiration, CNNs emulate the basic mechanism of the animal's visual cortex. Neurons in CNNs share weights unlike MLPs, where each neuron has a separate weight vector [22].

Using the weight-sharing strategy, neurons are able to perform convolutions on the data with the convolution filter formed by the weights. This is followed by a pooling operation that is a form of nonlinear subsampling that progressively reduces the spatial size of the representation, thus reducing the number of parameters and computation on the network [22].

After several layers of convolution and pooling, the size of the input matrix (size of the characteristics map) is reduced and more complex characteristics are extracted. Eventually, with a small enough feature map, the content is compressed into a one-dimensional vector and fed into a fully connected MLP for processing [22].

Normally, between the convolution layer and the pooling layer, a ReLu layer is applied to which an unsaturated activation function on one side is applied element by element, such as $f(x) = \max(0, x)$, limiting the lowest layer output values to zero [22].

The last layer of the fully connected MLP, seen as the output, is a loss layer that is used to specify how training on the network penalizes the deviation between the predicted and true labels [22].

Architecture of Convolutional Neural Networks

A CNN architecture is built on three components: convolutional layer, pooling layer, and a fully connected one. Convolutional layers use filters that examine a part of the image and extract resources from it. These resources are usually colors, shapes, and borders that define a specific image [23]. CNNs can have as many convolutional layers as needed. The more convolutional layers, the more resources are extracted.

One-dimensional convolution is an operation between a vector of weights $m \in R^m$ and a vector of inputs seen as a sequence $s \in R^s$. The vector m is the convolution filter. Concretely, we think of s as the input sentence and $s_i \in R$ is a unique characteristic value associated with the i-th word in the sentence. The idea behind the one-dimensional convolution is to take the scalar product of the vector m with each m-gram in the sentence to obtain another sequence c as shown in Eq. 1.

$$cj = m^t sj - m + 1 : j \qquad (1)$$

After the convolutional layers, we add pooling layers that are responsible for grouping the resource maps in an image [23], providing a reduction in dimensionality, which is produced by applying a single maximum or average of the values within a box produced by the convolutional layer. Thus, the fully connected layers, which are normal MLP networks, receive a smaller image than the one presented at CNN's entrance.

2.3   Image Classification

Image classification is mainly the method of isolating images in previously defined groups, where images related to the same group have significant affinities, as well as having significant distinctions with images from the other groups. The definition of image classification rules consists of extracting their characteristics.

For the construction of an automatic classifier to become viable, [24], p. 3 declare that "the rules for the classification of images must be expressed numerically, thus offering a technique developed or adapted to recognize patterns within these rules." Promptly for image processing, rules based on pixel intensity are selected to classify objects or regions of images.

Tomography Scans

Computed tomography (CT) is a diagnostic imaging exam consisting of an image representing a section or "slice" of the body. It is obtained through computer processing collected after exposing the body to a succession of X-rays. Its main method is to analyze an X-ray beam's attenuation as it travels through a segment of the body [25]. Among the characteristics of tomographic images, pixels, matrix, field of view, grayscale, and windows stand out. The pixel is the smallest point that can be obtained in an image. The greater the number of pixels in a matrix, the better its spatial resolution, which allows for better spatial differentiation between structures.

Functional Images

Functional imaging is the state of the art in medical diagnostic imaging. Thus, it was mentioned that "Magnetic Resonance has made a huge advance in this direction through ultra-fast obtaining techniques, which have made it possible to measure changes in the level of oxygen utilization resulting from the BOLD effect" [26], p. 1. There are two broad imaging methods in this line that are becoming evident: dynamic PET (positron emission tomography) and fMRI (functional magnetic resonance imaging). The first comes from nuclear medicine and the second from magnetic resonance. For the first, a device was developed to obtain images and, for the second, hardware and software techniques, taking advantage of the device's same base [27]. Functional images have been used mainly in research on brain functioning, being even able to provide information for surgical planning.

PET and Dynamic PET

According to Queirós [27] PET (positron emission tomography) is a diagnostic method developed with the purpose of schematizing the use of tissue glucose, transforming it into an efficient instrument for the detection of tumors. The isotope usually applied is fluoride linked to deoxyglucose, called FDG. Fluorine 18 is a positron emitter, a particle that rapidly interacts with the electrons in the medium (called annihilation), forming a pair of gamma photons, which pass from the point of annihilation in the same path, but in opposite directions. According to Phelps [28], "FDG is a matter metabolized by the cell indifferently, as well as glucose, due to its similarity. The main use is in the search for metabolic variations that indicate cancer or metastasis."

fMRI

Functional magnetic resonance imaging (fMRI) is a method of the special use of magnetic resonance imaging (MRI) qualified to detect changes in blood flow in reaction to neurological activity, a phenomenon called the BOLD effect [29]. This BOLD effect is based on the magnetic state of hemoglobin; that is, hemoglobin has the ability to present different magnetic states according to its oxygenation state. Thus, according to Pauling and Coryell [30], deoxyhemoglobin is paramagnetic (highlighted), that is, it magnetizes itself in the direction of the magnetic field to which it is presented, and oxyhemoglobin is diamagnetic (indented), and these magnetic properties have a direct effect on the strength of the signal detected in the neural regions that are active. It is possible to ascertain that an increase in the oxyhemoglobin agglomeration in the blood flow will generate an increase in the strength of the signal captured and that in an inverse situation, that is, in the presence of a greater agglomeration of deoxyhemoglobin, there will be a decrease in local strength due to the realignment of T2 and T2 *. This is because the events that start with the increase in electrical activity and articulate the neurovascular response change the magnetic resonance signal over time and generate the hemodynamic response function [31].

To study fMRI, it is essential to obtain one or more time series of functional information, acquired during the production of sensory or motor stimuli or during the production of paradigms, which are collections of cognitive tasks, and to obtain anatomical data that contain the areas of interest that exercise the function of structural reference for the visual concession of active functional areas. After this concession, the location and particularization of the brain regions activated by the stimuli is made. To this end, it is essential to perform image processing phases since this whole process is subject to the interference of types of artifacts that can modify the images obtained [32]. fMRI has been widely used in scientific research and much less in clinical follow-up. According to Langleben and Moriarty [33], "in many cases, fMRI is related to other non-invasive techniques such as electroencephalography (EEG) and near-infrared spectroscopy (NIRS)."

2.4 Tools and Techniques

This section shows the tools, frameworks, and techniques that were used during the development of the experiments.

TensorFlow

TensorFlow is a machine learning system that operates on a large scale and in heterogeneous environments. TensorFlow uses data flow graphs to represent computing, shared state, and operations that change that state. It maps the nodes of a data flow graph on many machines in a cluster and, within one machine, on various computing devices, including multicore CPUs, general-purpose GPUs, and custom ASICs, known as Tensor Processing Units (TPUs). This architecture offers application developer flexibility: while in previous "parameter server" designs, shared state management is integrated into the system, TensorFlow allows developers to experiment with new optimizations and training algorithms. TensorFlow is compatible with a variety of applications, with a focus on training and inference in deep neural networks [34].

Keras

It is the high-level API of TensorFlow for creating and training deep learning models. It is used for rapid prototyping, cutting-edge research, and production, with the following advantages: It has a simple and consistent interface optimized for common use cases, and Keras modular and composite models are made by connecting configurable elements, with few restrictions [35].

NVIDIA CuDNN

It is a GPU-accelerated library of primitives for deep neural networks. CuDNN provides highly tuned implementations for standard routines, such as forward and backward convolution layers, pooling, normalization, and activation. CuDNN accelerates the widely used deep learning frameworks, including Caffe2, Keras, PyTorch, and TensorFlow [36].

Transfer Learning

It is the reuse of a pre-trained model in a new problem. The models used were trained with weights from ImageNet, which has millions of natural images in several categories. Training such a large mass of data might be very expensive, taking many days in advanced GPUs, then expending too much energy, consequently, money.

The purpose of transfer learning is to take advantage of all this knowledge that was generated using these natural images, whether in the most initial layers, such as the recognition of edges, curves, and colors, or in the innermost convolutional layers where they learn the texture, smooth. The more specialized layers are not interesting for learning, so only the initial layers are used that recognize basic details that theoretically every image has.

Callbacks

A callback is an object that can perform actions at various stages of the training (e.g., at the beginning or at the end of a season, before or after a single batch, etc.) [34]. Callbacks can be used to:

- Record TensorBoard records after each training batch to monitor your metrics
- Periodically save your model to disk
- Stop early
- Get a view of a model's internal states and statistics during the training

3   Experiments

This section details how the experiments were carried out, what equipment was used during the process, and how the data were obtained and treated for the training of the neural networks, in addition to the metrics used to evaluate the architectures.

3.1   Environment

All experiments were performed using a machine with the following configurations:

- CPU Intel Core i7 3770 @ 3.40GHz
- Memory 8GB Dual-Channel DDR3 @ 798MHz (10-10-10-30)
- GPU NVIDIA GeForce GTX 960 4GB (ZOTAC International)
- SSD 240GB KINGSTON (SATA-3)

3.2   Dataset

A Kaggle dataset called the SARS-COV-2 Ct-Scan Dataset provided by Soares et al. [9] was used, which united the researchers to the study data, and this dataset was used for the first stage of this project. The dataset consists of 2477 computed tomography images used to detect lung diseases, 1250 of which tested positive for COVID-19 and 1227 had a negative diagnosis. Figures 3 and 4 show a sample of the dataset with images of people with positive and negative diagnoses, respectively.

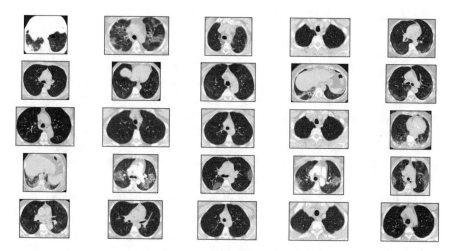

Fig. 3 Standardized COVID-19 images, [9]

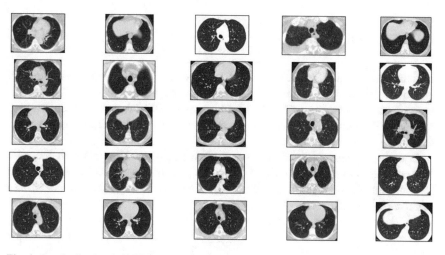

Fig. 4 Standardized non-COVID-19 images, [9]

The dataset has images with different resolutions; due to this, all images were converted to 128 × 128 resolution. In addition, the images were standardized and the validation sets were separated using ImageDataGenerator; this technique also makes image augmentation which generates new images while separating and normalizing. The following operations were also performed in order to avoid overfitting the neural network: rotation range = 10, width shift range = 0.2, height shift ranged = 0.2, zoom range = 0.1, horizontal flip = True, and vertical flip = True.

3.3   Definition of Models

The model was created using "include_top = false" to remove the specific piece of natural images when starting with the weights of ImageNet. Soon after, a pooling is added to reduce the size, and some dense layers are added. In dense layers, "relu" was used as activation and "he_normal" as kernel weight. Dropout with 0.3 was also used in each dense layer, and lastly, softmax to be able to separate the classes.

Adam was used with a learning rate of 0.002, the loss function as categorical_ crossentropy, and accuracy as a metric. Soon after, the model is executed using cross-validation with $k = 5$ and fit_generator with 50 periods for each of the architectures.

The following callbacks were also used to perform actions at various training stages:

- ModelCheckPoint: to save the model that has the best loss during the training
- EarlyStop to stop training if the network stops learning
- ReduceLROnPlateau to decrease the learning rate if the loss of validation does not change

3.4   Evaluation Metrics

Initially, the meaning of true positives, true negatives, false positives, and false negatives was defined. True positives and true negatives are correct classifications, that is, classifying COVID-19 and non-COVID-19 correctly. In contrast, false positives and false negatives are wrong classifications. Thus, metrics were defined.

Equation 2 presents the first metric, called precision. This metric is the ratio between true positives and true positive plus false positives. Thus, a low precision indicates that the number of correct classifications is too low, or the number of false positives too is high.

$$P = \frac{TP}{TP + FP} \tag{2}$$

The next metric is the accuracy, as presented in Eq. 3, which is the percentage of correct classifications. A low accuracy could indicate that the number of wrong classifications (false positives and false negatives) is high.

$$A = \frac{TP + TN}{TP + FP + TN + FN} \tag{3}$$

The recall presented in Eq. 4 is the ratio between the true positives and true positives plus false positives. This metric indicates that the algorithm is performing well in classifying true positives. However, if this metric is low, it can mean that a high number of misclassifications is going on. Thus, this metric is essential to minimize the number of false negatives, which can produce the patient's worst scenario.

$$R = \frac{TP}{TP + FN} \tag{4}$$

Finally, the F1 score presented in Eq. 5 is the harmonic mean between precision and recall. In this context, F1 ends up being a big picture of the performance because precision takes into account false positives, and recall takes into account false negatives. Thus, F1 score gives an idea of whether the classifying algorithm is providing too many incorrect classifications.

$$F1score = \frac{2 \times precision \times recall}{precision + recall} \tag{5}$$

4 Results

Table 1 shows the best results for the accuracy and loss of the 11 CNNs according to the metrics presented above. In the same table, the average of all metrics used during training is also presented. As can be seen, DenseNet169 achieves the best precision, recall, F1_score, and accuracy of 96.3%, 96.4%, 96.4%, and 96.4%, respectively. In addition, DenseNet169 presented the best fold accuracy with 96.9%. MobileNetV2 had the worst results.

Considering that DenseNet169, VGG16, Xception, and VGG19 obtained the best results, these will be detailed below. DenseNet169 presented the best result. Figure 5 shows its average confusion matrix, in which we can observe that DenseNet169 was incorrect in only 18 cases; however, 12 cases gave false positives for COVID-19. On the other hand, the notable fact is that only six cases of COVID-19 were erroneously classified as non-COVID-19. Below, we detail the results of DenseNet169, showing the confusion matrices and training curves.

Figures 6, 7, 8, 9, and 10 show the accuracy and loss fold. As can be seen, both curves demonstrate the expected behavior, which is to achieve a value of one (100%)

Table 1 Result of 11 CNNs: accuracy, precision, recall, F1_score, accuracy fold, and loss fold

Model	Best fold accuracy	Best fold loss	Average precision	Average recall	Average F1_score	Average accuracy	Average loss
DenseNet169	*0.969*	*0.097*	*0.963*	*0.964*	*0.964*	*0.964*	*0.097*
VGG16	0.957	0.108	0.952	0.951	0.951	0.951	0.134
Xception	0.965	0.126	0.951	0.951	0.951	0.951	0.150
VGG19	0.967	0.109	0.947	0.946	0.946	0.946	0.134
DenseNet201	0.961	0.119	0.949	0.948	0.948	0.948	0.138
DenseNet121	0.957	0.153	0.946	0.947	0.947	0.947	0.155
ResNet50	0.959	0.108	0.943	0.942	0.942	0.943	0.149
ResNet50V2	0.949	0.134	0.940	0.940	0.939	0.939	0.163
InceptionV3	0.957	0.140	0.940	0.938	0.939	0.939	0.162
MobileNet	0.955	0.112	0.935	0.934	0.934	0.934	0.165
MobileNetV2	0.752	0.576	0.634	0.613	0.588	0.617	0.644

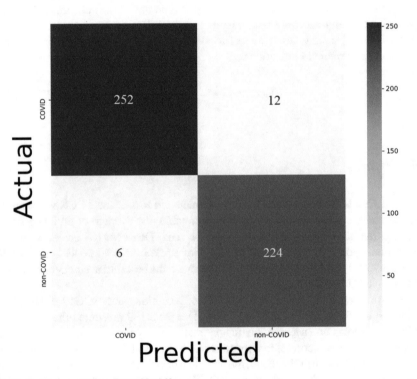

Fig. 5 Confusion matrix – DenseNet169

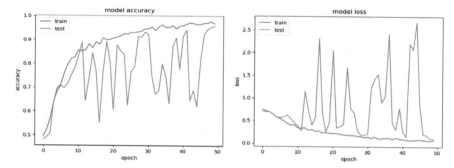

Fig. 6 Accuracy and loss fold 1 – DenseNet169

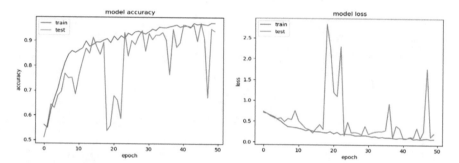

Fig. 7 Accuracy and loss fold 2 – DenseNet169

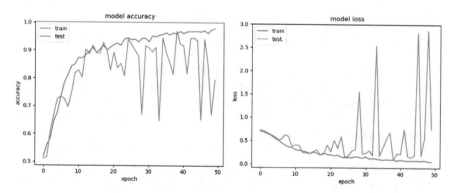

Fig. 8 Accuracy and loss fold 3 – DenseNet169

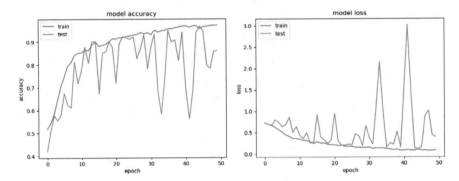

Fig. 9 Accuracy and loss fold 4 – DenseNet169

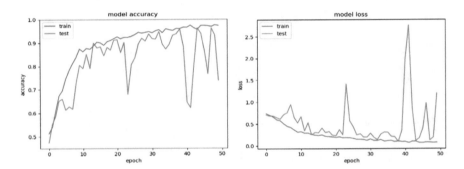

Fig. 10 Accuracy and loss fold 5 – DenseNet169

in precision and decrease to zero (0%) in the loss. In other words, the curves tend to converge.

VGG16 also achieved good results. Figure 11 shows its average confusion matrix, in which we can observe that VGG16 was incorrect in 24 cases; however, 12 cases gave false positive for COVID-19, and 12 cases of COVID-19 were wrongly classified as non-COVID-19. Below, we detail the results of VGG16, showing the confusion matrices and training curves.

Figures 12, 13, 14, 15, and 16 show the accuracy and loss fold achieved by VGG16. As can be seen, both curves demonstrate the expected behavior, which is to achieve a value of one (100%) in precision and decrease to zero (0%) in the loss. In other words, the curves tend to converge.

Figure 17 shows the average confusion matrix of the Xception, which also achieved good results, in which we can observe that Xception was incorrect in 24 cases; however, 14 cases gave false positives for COVID-19. On the other hand, only ten cases of COVID-19 were erroneously classified as non-COVID-19. Below,

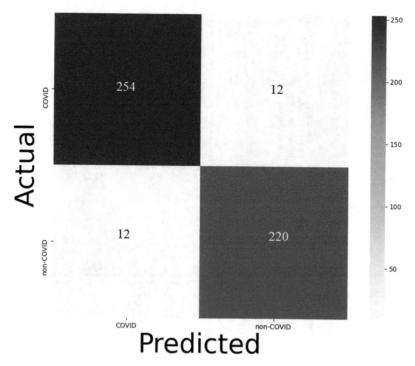

Fig. 11 Confusion matrix – VGG16

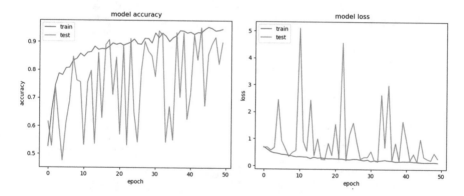

Fig. 12 Accuracy and loss fold 1 – VGG16

we detail the results of Xception, showing the confusion matrices and training curves.

Figures 18, 19, 20, 21, and 22 show the accuracy and loss fold achieved by Xception. As can be seen, both curves demonstrate the expected behavior, which is to achieve a value of one (100%) in precision and decrease to zero (0%) in the loss. In other words, the curves tend to converge.

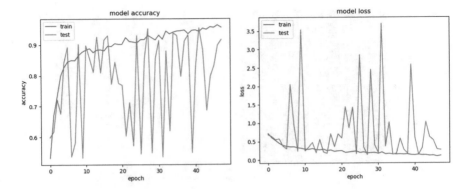

Fig. 13 Accuracy and loss fold 2 – VGG16

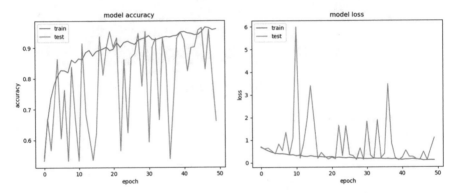

Fig. 14 Accuracy and loss fold 3 – VGG16

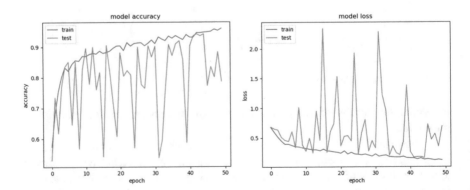

Fig. 15 Accuracy and loss fold 4 – VGG16

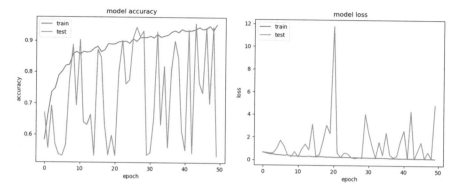

Fig. 16 Accuracy and loss fold 5 – VGG16

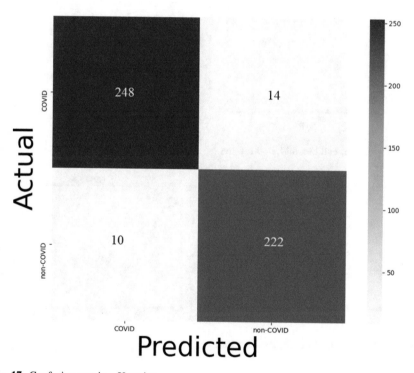

Fig. 17 Confusion matrix – Xception

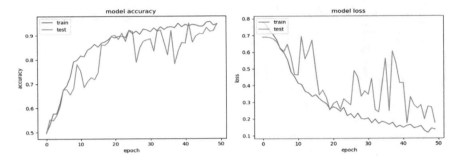

Fig. 18 Accuracy and loss fold 1 – Xception

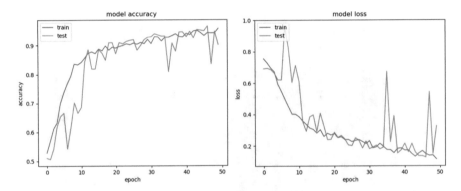

Fig. 19 Accuracy and loss fold 2 – Xception

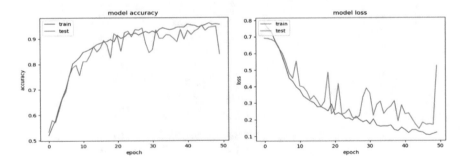

Fig. 20 Accuracy and loss fold 3 – Xception

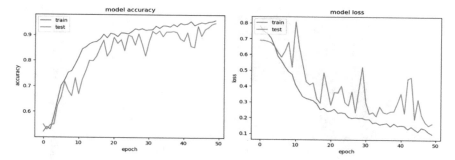

Fig. 21 Accuracy and loss fold 4 – Xception

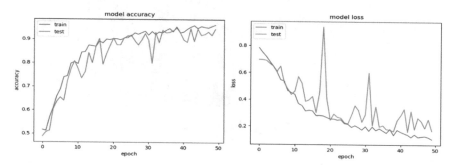

Fig. 22 Accuracy and loss fold 5 – Xception

Finally, Fig. 23 shows the average confusion matrix of VGG19, which also achieved good results, in which we can observe that VGG19 was incorrect in 27 cases; however, 12 cases gave false positives for COVID-19, and 15 cases of COVID-19 were erroneously classified as non-COVID-19. Below, we detail the results of VGG19, showing the confusion matrices and training curves.

Figures 24, 25, 26, 27, and 28 show the accuracy and loss fold achieved by VGG19. As can be seen, both curves demonstrate the expected behavior, which is to achieve a value of one (100%) in precision and decrease to zero (0%) in the loss. In other words, the curves tend to converge.

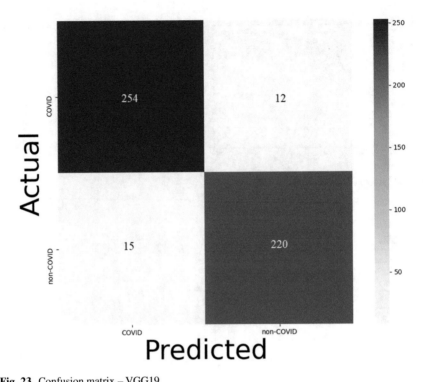

Fig. 23 Confusion matrix – VGG19

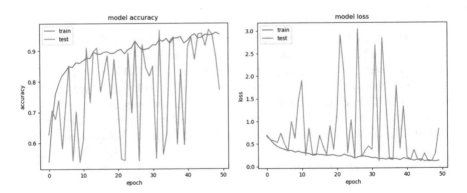

Fig. 24 Accuracy and loss fold 1 – VGG19

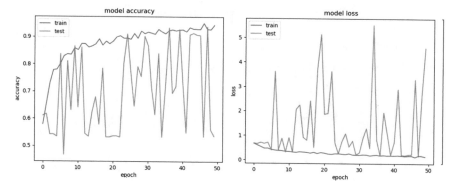

Fig. 25 Accuracy and loss fold 2 – VGG19

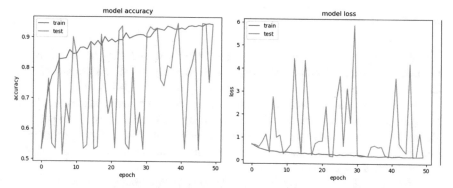

Fig. 26 Accuracy and loss fold 3 – VGG19

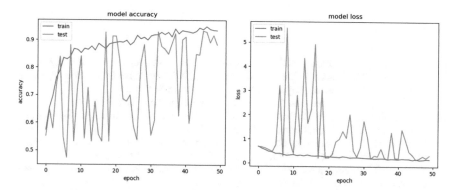

Fig. 27 Accuracy and loss fold 4 – VGG19

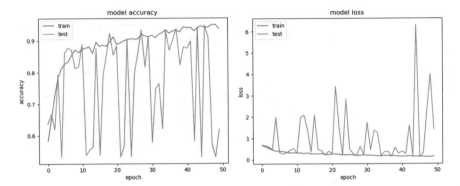

Fig. 28 Accuracy and loss fold 5 – VGG19

5   Conclusion

This research presented the performance of 11 CNNs trained using transfer learning in the classification of tomographic images, particularly of patients diagnosed with COVID-19 and patients who had a negative diagnosis for COVID-19.

The initial results showed that DenseNet169 achieved better results in all the averages of the metrics used, in addition to having achieved the best result per fold. DenseNet169 classified most cases of patients with COVID-19, missing on average only 6 cases. Also, the accuracy and loss curves showed an expected behavior toward 100% accuracy and 0% loss.

In future work, these 11 CNNs will be tested on other image datasets.

References

1. R.S. Esteves, A.C. Lorena, M.Z. Nascimento, Aplicação de técnicas de Aprendizado de Máquina na Classificação de Imagens Mamográficas, 2009. Available in: http://ic.ufabc.edu. br/II_SIC_UFABC/resumos/paper_5_150.pdf. Accessed 15 Sept 2020
2. B. Wolf, C. Scholze, "Medicine 4.0". Current directions in. Biom. Eng. **3** (2017)
3. H.P. Schnurr, D. Aronsky, D. Wenke, *MEDICINE 4.0—Interplay of Intelligent Systems and Medical Experts* (Springer, Cham, 2018), pp. 51–63
4. M. Monard, J. Baranauskas, Conceitos sobre aprendizado de máquinas. Em Sistemas Inteligentes: Fundamentos e Aplicações, Cap. 4, Editora Manole, 512 P (2003)
5. Y. Xin et al., Aprendizado de máquina e métodos de aprendizado profundo para segurança cibernética. Acesso IEEE **6**, 35365–35381 (2018)
6. D. Sarkar, R. Bali, T. Ghosh, *Transfer Learning with Python* (Pack Publishing, 2018)
7. World Health Organization, Situation Report – 158 2020. https://www.who.int/publications/m/ item/weekly-update-on-covid-19%2D%2D-23-october. Accessed 23 Oct 2020
8. Deep Learning 0.1. Convolutional Neural Networks (LeNet) – DeepLearning 0.1 documentation. LISA Lab. Accessed 15 Sept 2020
9. E. Soares et al., SARS-CoV-2 CT-scan dataset: A large dataset of real patients CT scans for SARS-CoV-2 identification. medRxiv (2020)

10. M.E.H. Chowdhury, T. Rahman, A. Khandakar, R. Mazhar, M.A. Kadir, Z.B. Mahbub, K.R. Islam, M.S. Khan, A. Iqbal, N. Al-Emadi, M.B.I. Reaz, M.T. Islam, Can AI help in screening Viral and COVID-19 pneumonia? IEEE Access **8**, 132665–132676 (2020)
11. T. Mitchell, *Machine Learning* (McGraw Hill, 1997)
12. G.L. Libralão, R.M. Oshiro, A.V. Netto, A.P.L.F. Carvalho, M.C.F. Oliveira, Técnicas de Aprendizado de Máquina para análise de imagens oftalmológicas (2003). Available in: http://www.lbd.dcc.ufmg.br/colecoes/wim/2003/003.pdf. Accessed 15 Sept 2020
13. T.W. Rauber, Redes neurais artificiais (2005). Available in: http://www.riopomba.ifsudestemg.edu.br/dcc/dcc/materiais/1926024727_reconhe cimento-de-caracter2.pdf. Accessed 10 Sept 2020
14. W.S. McCulloch, W. Pitts, A logical calculus of the ideas immanent in nervous activity. Bull Math. Biophys. **5**, 115–133 (1943)., Reprinted in [Anderson and Rosenfeld, 1988].
15. F. Rosenblatt, The perceptron: A probabilistic model for information storage and organization in the brain. Psychol. Rev. **65**, 386–408 (1958)
16. B. Widrow, M.E. Hoff, Adaptive switching circuits, in *1960 WESCON Convention Record, New York,* (1960)
17. P.S. Prampero, Combinação de classificadores para reconhecimento de padrões, Dissertação (Mestrado), Instituto de Ciências Matemáticas e Computação – USP, São Carlos – SP, p 84., 12 out. 2017 1998
18. C.Y. Tatibana, D.Y. Kaetsu, Uma introdução às redes neurais 2000. Available in: http://www.din.uem.br/ia/neurais. Accessed 10 Sept 2020
19. L.G. Perez et al., Training an artificial neural network to discriminate between magnetizing inrush and internal faults. IEEE Trans. Power Delivery **9**(1), 434–441 (1994)
20. Carvalho, A. C. P. L. F. Notas de aula: SCE5809 – Redes Neurais, Grupo de Inteligência Computacional, Pós-graduação em Computação, ICMC–USP, São Carlos-SP (1999)
21. Y. Lecun, Y. Bengio, G. Hinton, Deep learning. Nature **521**(7553), 436–444 (2015)
22. F.S. Melo, Extração de relações a partir de dados não estruturados baseada em deep learning e supervisão distante, Dissertação (Mestrado) – Curso de Ciência da Computação, Universidade Federal de Sergipe, São Cristóvão 2018
23. T. Beysolow, *Introduction to Deep Learning Using R: A Step-by-Step Guide to Learning and Implementing Deep Learning Models Using R* (Apress, 2017)
24. F.J. Silva, R.M.S. Julia, A. Barcelos, Diagnóstico de Imagens de Ressonância Magnética quanto a presença de Esclerose Múltipla por meio de Redes Neurais Artificiais. BDBComp, Minas Gerais, **1**(1) 2012. Available in: http://www.lbd.dcc.ufmg.br/colecoes/enia/2012/008.pdf. Accessed 15 Sept 2020
25. E. Amaro, H. Yamashita, Aspectos básicos de tomografia computadorizada e ressonância magnética. Braz. J. Psychiatry **23**, 2–3 (2001)
26. E. Tinois, Imagem Funcional PET E fMRI. Tecnologia para a Saúde V, Campinas, **1**(1) outubro (2005). Available in: http://www.multiciencia.unicamp.br/artigos_05/a_02_05.pdf. Accessed 15 Sept 2020
27. G.C. Queirós, Análise Computacional de Imagens de Ressonância Magnética Funcional (2011). Available in: https://web.fe.up.pt/~tavares/downloads/publications/relatorios/Monografia_Gabriela_Queiros.pdf. Accessed 15 Sept 2020
28. M.E. Phelps, *PET Molecular Imaging and Its Biological Applications* (Springer Verlag, New York, 2004). Available in: http://www.springer.com/la/book/9780387403595. Accessed 15 Sept 2020
29. S. Ogawa, T.M. Lee, A.R. Kay, D.W. Tank. Brain magnetic resonance imaging with contrast dependent on blood oxygenation. Proc. Natl. Acad. Sci. U. S. A. 87(24), 9868–9872 (1990)
30. L. Pauling, C.D. Coryell, The magnetic properties and structure of hemoglobin, oxyhemoglobin and carbonmonoxyhemoglobin. Proc. Natl. Acad. Sci. **22**, 210–216 (1936)
31. S.H. Faro, F.B. Mohamed, *BOLD fMRI – A Guide to Functional Imaging* (Springer Science, New York, 2010)

32. E. Formisano, F. Salle, R. Goebel, Fundamentals of data analysis methods in functional MRI, in *Advanced Image Processing in Magnetic Resonance Imaging*, ed. by L. Leandini, V. Positano, M. F. Santarelli, (Taylor & Francis Group, LLC, 2005), pp. 481–500
33. D.D. Langleben, J.C. Moriarty, Using brain imaging for lie detection: where science, law, and policy collide. Psychology, Public Policy, and Law **19**(2, 222) (2013). Available in: https://pt.wikipedia.org/wiki/Imagem_por_ress%C3%B4nancia_magn%C3%A9tica_funcional#cite_note-1. Accessed 15 Sept 2020
34. M. Abadi et al., Tensorflow: a system for large-scale machine learning, in *12th {USENIX} Symposium on Operating Systems Design and Implementation ({OSDI} 16)*, (2016), pp. 265–283
35. Keras. About. 2020. Available in: https://keras.io/, Accessed 10 Sept 2020
36. S. Chetlur et al, cudnn: Efficient primitives for deep learning. arXiv preprint arXiv:1410.0759, (2014)

Challenges in Processing Medical Images in Mobile Devices

Mariela Curiel and Leonardo Flórez-Valencia

1 Introduction

The processing of medical images helps health professionals make decisions for the diagnosis and treatment of patients. However, medical images pose certain challenges for healthcare institutions, among them ubiquitous access and efficient processing.

Medical images must be easily accessed anywhere and anytime. Ubiquity can be achieved by mobile devices, specifically smartphones. Nowadays, there are about 5.190 million unique users on mobile devices. The figure has grown annually and, compared to 2019, had a constant growth of 2.4% (Shum, 2020). Smartphones have been widely used in health matters [e.g., [1–4]], especially in image analysis [5].

Although some algorithms for image processing may be entirely executed in a mobile device, in other cases, due to the volume of images, the processing requires substantial amounts of resources. It must be carried out in higher capacity platforms, such as workstations. Even in these platforms, if the algorithms are centralized, they could not scale well with the image size. This leads us to think of distributed alternatives to perform this processing without losing the ubiquity. The use of mobile devices would help us with both purposes. Mobile devices would bring additional advantages, such as the reduction in costs and space in computing resources for healthcare institutions as well as the reach to places and communities with low penetration by other types of computer systems.

Recent advances in mobile devices' computing capabilities allow them to handle computationally intensive applications, whereupon they can be considered the computing platforms of the future [6]. Improvements consist of multi-core CPUs, several gigabytes of RAM, and communication through several wireless networking

M. Curiel (✉) · L. Flórez-Valencia
Pontificia Universidad Javeriana, Bogotá, Colombia
e-mail: mcuriel@javeriana.edu.co; florez-l@javeriana.edu.co

© Springer Nature Switzerland AG 2022
P. Johri et al. (eds.), *Trends and Advancements of Image Processing and its Applications*, EAI/Springer Innovations in Communication and Computing,
https://doi.org/10.1007/978-3-030-75945-2_2

technologies. Some examples of the use of smartphones as computing platforms can be found in [7–9].

In the same address, several researchers have worked in incorporating mobile devices into the grid (e.g., [10–12]). Particularly, a *mobile grid* is a specific grid where users can connect using their mobile devices (smartphones, tablets, etc.) to either obtain access to all of the grid's resources or place their mobile devices as a computing platform.

We are currently testing the feasibility of this technology for image processing. The use of mobile grids and, particularly, the use of mobile devices as execution platforms bring the following challenges:

(a) Choose the distributed execution platform that adapts to the requirements of the research project. The executing platform requirements are open-source code and support for C++ programming language and image processing libraries (e.g., ITK), with complete and adequate documentation.
(b) Support the programmer's task. Moving existing algorithms to mobile devices or developing new algorithms requires new knowledge of these platforms. In this sense, one of our objectives is to provide transparency for the programmer, i.e., the execution of processing image algorithms in mobile devices should not require changes in the source code.
(c) Proper use of the devices. Mobility and battery life, coupled with intermittence, are some of the challenges that should be solved to utilize mobile devices as execution platforms fully.

This chapter presents the lessons learned trying to solve these challenges. The rest of this chapter is structured as follows: Section 2 details the research's main concepts. Section 3 exposes related work. Sections 4, 5, and 6 have responses to the previously described challenges. Lastly, in Sect. 7, learned lessons and future work are presented.

2 Background

2.1 Healthcare and Image Processing

Telemedicine is the care of patients remotely, using telecommunication and information technologies to provide healthcare at a distance. Thanks to telemedicine, rural and more distant communities can have access to health services.

Image processing is the name given a wide set of computational techniques for analyzing, enhancing, compressing, and reconstructing images. The techniques' output can be either an image or a set of features or parameters related to the image. One important subset of such techniques is composed of filters. Filters are algorithms that take images as inputs and perform operations over them, such as noise removal and image enhancement. The output is, usually, another image that highlights or deemphasizes certain characteristics of the input image. In the spatial domain, most filters operate over each pixel independently, modifying its value given an operator matrix or kernel and the neighboring pixel's values. On the other hand, the frequency domain operates with the rate at which the pixels change over the spatial domain.

2.2 Distributed Computing

Grid computing involves the aggregation of heterogeneous resources, geographically dispersed and from different organizations, to solve computationally complex problems [13].

In [11] a mobile grid is defined as a grid that includes at least a mobile device. Some researchers strictly define ad-hoc grids as grid environments without fixed infrastructures, i.e., all their components are mobile [10, 14].

Voluntary computing is a type of distributed computing, where volunteers' computing or storage resources are used [15]. Volunteers are usually ordinary people, probably geographically dispersed, that have computing resources (desktops, laptops, tablets, smartphones, etc.) with an Internet connection. Volunteers are not just people; organizations can also provide their computing resources available.

Cloud computing, widely recognized as the next-generation computing infrastructure, offers on-demand computer, data storage, and software as a service without the user's direct active management.

The explosion of mobile applications and the support of cloud computing for mobile users' services promoted mobile cloud computing's birth as the integration of cloud computing into the mobile environment. The work presented in [16] remarks that mobile cloud computing allows the execution of applications on resource providers external to the mobile device [17], i.e., in the cloud.

Mateos et al. [18] propose the use of wrappers to gridify applications (i.e., to execute applications in a grid). This is said to be a coarse-grained gridification technique where applications are taken in their binary form along with some configuration parameters provided for the users. The executable code is wrapped with a software component that isolates the complex details of the underlying grid.

The cross-compilation uses a compiler to generate executable code for a platform different from where compiling occurs.

3   Related Work

Mobile health (mHealth) describes the use of portable electronic devices with software applications to provide health services and manage patient information [19]. In this section, several works related to smartphones' use in medical image processing will be exposed. For a more comprehensive review of the state of the art, refer to [5].

In [20], the authors developed a client-server application for efficient medical image segmentation. The application works in three steps: (1) First, some selected medical images are uploaded from a server computer to the picture gallery of a smartphone. (2) Second, a medical image to be segmented is selected from the picture gallery, and it is sent to the server with some input parameters. (3) Finally, the split-Bregman iterative method is applied at the server side, whereupon the segmented image is automatically sent to the smartphone.

The work presented in [21] describes an interactive method for segmenting and visualizing medical images on mobile devices. The system, so-called ISVS_M-2 (Interactive Segmentation and Visualization System on Mobile Devices), works on workstations and a wide range of mobile devices using the client/server paradigm. The application comprises three modules: (a) a segmentation module that works on a server, (b) several commutation modules that run on both the server and mobile devices, and (c) interactive and visualization modules on mobile devices. The interactive and visualization modules allow the expert to visualize the internal information of a model, as well as to refine its segmentation interactively according to their experience.

Vaish et al. [22] support computer-aided diagnosis (CAD) using smartphones. The authors developed an Android OS application that automatically detects kidney abnormalities in B-mode ultrasound images. This application allows smartphones to obtain kidney images from an ultrasound scanner, process them, and give the image's diagnostic result.

After surgery, it may be difficult for some patients to visit the medical consultation personally repeatedly. To provide some help to such patients, smartphone-based applications have been designed, which allow doctors to monitor the post-surgery wounds. An example is the work presented in [23]. The author implemented an algorithm on IOS devices for deforming the surgery site image to help monitor surgery site infection (SSI) risk.

All the previously mentioned works have in common that image management is done on a single device (or executing platform) or shared between a mobile device and a server. The following articles propose efficient and scalable solutions that take advantage of other technologies, such as the existence of multiple cores in current computers and cloud computing technology. Cloud technologies provide not only efficiency but ubiquity.

The work described in [24] proposes an algorithm to reduce the computational complexity of compressed sensing-based magnetic resonance image (MRI) reconstruction algorithms. To take advantage of the multi-core architecture, the authors break down the CS-based MRI reconstruction problem into four independent sub-problems. The sub-problems are assigned to four cores of the CPU, where they are solved simultaneously, accelerating the MRI reconstruction. Compared to the original problem, each sub-problem has lower computational complexity, and the proposed MRI reconstruction is speeded at least four times up.

Zhuang et al. [25] propose cloud services to increase the performance of an algorithm for image retrieval. The article states that most of the algorithms for medical image indexing and similarity query processing in high-dimensional spaces have been designed for a single computer, i.e., they are based on a centralized approach. Unfortunately, centralized approaches do not scale up well in the presence of large data volume. As a solution, the article describes MIRC, a robust and efficient, content-based method for retrieving large medical images in a mobile cloud computing environment.

Vazhenin D. describes a web-based system that provides medical information that can be retrieved using a wide variety of devices (workstations, smartphones, etc.) [26]. The client side of the application interacts with the end-user and allows the visualization of the rendered information. On the other hand, the server performs medical image archive management, preprocessing, and rendering images in the cloud.

Liu et al. [27] propose the iMAGE cloud, a software as a service (SaaS) cloud. The article describes a three-tier hybrid cloud that provides medical imaging services in the smart city of Wuxi, China. The layers are described below: In the first layer, the medical images and EMR (electronic medical records) data are received and integrated through the hybrid regional health network. Traditional and advanced image processing functions reside in the second layer, where high-performance cloud units are computed. Finally, the image processing results are delivered to regional users using virtual desktop infrastructure (VDI) technology.

4 The Platform for Distributed Execution

In order to address the execution of algorithms on mobile devices, we decided to adopt an existing grid instead of building a system from scratch, as the works described in [9, 28, 29]. The reason is that an existing grid system provides functionalities already implemented that save time, such as resource discovery and scheduling.

This section describes the process to choose a suitable execution platform with mobile devices, the problems faced, and how they were solved.

4.1 Selection

At the beginning of this research in 2013–2014, a bibliographic review was carried out to identify all the possible implementations of mobile grids. Many of them were discarded because the systems were not publicly available. Such was the case of Akogrimo [30], MiPEG [31], MORE [32], Mobile OGSI.NET [33], and MADAM [34]. The tools that were found in the literacy and whose source code was available were BOINC [35], Ibis [36], MoGrid [37], and OurGrid [38]. According to the research project, the work of [39] describes the methodology to evaluate these four systems and the attributes we expected from them. These attributes are:

- *Execution on mobile devices*: Our research project's goal was to use mobile devices as processing platforms as they provide ubiquity and can be used in hard-to-reach places.
- *Supported programming languages*: The execution of tasks in C++ was needed since most of the libraries and frameworks for the processing and analysis of medical images are written in this language. The execution of other programming languages was also desirable. It is important to mention that currently, other systems also allow the execution of applications on mobile devices (e.g., [9, 40–42]). However, they only support the execution of applications written in JAVA.
- *Project maintainability*: It was important to know the state of the grid project at the time of the evaluation, i.e., if the project was still under development or achieved a distribution version to the public. These characteristics guarantee complete software, free of errors, and the existence of a community that can give support in case of problems. A project that has been abandoned does not present these advantages, nor will it be incorporating the technological innovations that continually come out.

Another desirable requirement to meet in the medium term is to develop our system on an ad-hoc grid that can accommodate nearby mobile devices in case of emergencies or disasters.

Even though the BOINC model is not ad hoc, we chose this platform because it met the fundamental requirements in the project and especially because BOINC was the only system that provided an Android-based client to execute tasks. Other systems for mobile volunteer computing that exist today are CrowdLab [43], Green Energy Mobile Cloud (GEMCloud) [44], computing while charging (CWC) [9], and FemtoCloud [45]. The next section describes BOINC.

4.2 BOINC

BOINC (Berkeley Open Infrastructure for Network Computing) is an open grid framework that has been used for voluntary grid computing projects. A volunteer makes unused compute cycles on their devices (personal computers, laptops, or mobile devices) available to BOINC. Various research projects involving distributed computing take advantage of these unused cycles [35]. By incorporating mobile devices as computing nodes, BOINC can be classified as a mobile grid. The following paragraphs explain the main concepts of BOINC, which were taken from https://boinc.n-helix.com/trac/wiki/BasicConcepts.

A BOINC *project* corresponds to an organization or research group that must do volunteer or distributed computing. The project is identified by a master URL, which is the home page of its web site. Participants register with projects, and a project can involve one or more applications.

An *application* includes several executable codes for different platforms (Windows, Linux/x86, Mac OS/X, etc.), a set of workunits, and results. An application may be composed of a sequence of versions. An *application version* is a version compiled for a particular platform. A *platform* is a compilation target that typically includes a CPU architecture and an operating system. The *workunit* is the computation to be performed, i.e., a "job" or a "task" (we use the term task throughout this document). In some cases, there may be several instances or "replicas" of a given workunit. The workunit may include input files and various attributes, such as resource requirements and deadlines. A *result* describes an instance of a computation that can be uninitiated, in progress, or completed. Each result is associated with a workunit.

The *server* of a BOINC project stores description of applications, platforms, versions, workunits, results, accounts, teams, etc. Volunteers who want to be part of a BOINC project must visit the project's website to download the *BOINC Client*.

The following steps describe the life cycle of a workunit:

1. A Work Generator creates a workunit along with their parameters and input files.
2. BOINC creates one or more instances of the workunits and dispatches them to different hosts.
3. In the host, a client program downloads the input files and executes the workunit.
4. Upon completion of the workunit, the host reports its completion to the server.
5. In the server, the *Validator* checks the output files by probably comparing replicas of the same result.
6. The *Assimilator* program handles valid results (e.g., by inserting them in a separate database).
7. When all instances have been completed, the File Deleter removes the input and output files.

Task Scheduling The task scheduling process in BOINC is associated with its volunteer computing model. In this model, both the client (or volunteer) and the server execute the scheduling algorithm.

When a volunteer is registered in a project, the scheduling policy executed on the client periodically makes requests to the task scheduler running on the server. The server task scheduler responds to requests sent by volunteers. The selection of the workunits that will be sent to the volunteers is done through a policy based on a scoring function. The scoring function $P(N,W)$ looks to send to the client N the workunits W for which the value of the function $P(N,W)$ is a maximum. This function, which is used in BOINC [46] as a default, combines parameters related to (i) the expected amount of floating-point operations per second (FLOPS) of the application version available for client N, (ii) memory and storage requirements of the application, (iii) the number of replicas to be sent, and (iv) homogeneous redundancy class to which N belongs [47]. The use of replicas is explained below.

The assignment of tasks to volunteers is as follows: The process starts when a client requests tasks to run. The request is received by the BOINC SERVER, which checks if the volunteer user has enough privileges to access the project for which it is requesting tasks. After performing the necessary checks, the server passes the request to the scheduler by calling the *send_work_score* routine, which in turn implements the scoring function described above. After selecting the tasks to be sent, the *add_results_to_reply* routine adds the tasks into the *assigned_tasks* array. The group of selected tasks is returned to the BOINC SERVER, where they are formatted into an HTTP/XML message that is sent back to the client to be executed. When the tasks are finished, the client returns the result to the server and asks for new tasks.

Replication In volunteer computing, anyone with a device (smartphone, tablets, or PCs) can be a participant, and it is impossible to be sure that everyone is reliable or well-intentioned. Additionally, since BOINC is open source and easily accessible on the Internet, there is a risk of receiving invalid results from one or more clients. To mitigate this risk, BOINC implements a replication mechanism in which the same *workunit* is sent to several clients. Then, through a quorum system, the server verifies that the result returned by a volunteer matches with the results coming from other volunteers for the same *workunit*. Each project enables this mechanism by indicating the number of replicas to be generated and the number of results required in the quorum to mark a *workunit* as completed. There is also adaptive replication, which consists of determining whether a device is reliable or not depending on the amount of consecutive satisfactory results that it returns to the server and the time it takes to execute the *workunit*. Adaptive replication is recommended by BOINC because it tries to reduce the number of replicas generated over time.

5 Application Execution and Transparency for the Programmer

Although BOINC allows us the execution of tasks in Android OS, it was necessary to make changes in some of its components to avoid the modification of user programs, compile the existing libraries, and program the parallel processing of images. The next paragraphs describe the experience.

5.1 The Wrapper for Android OS

Running a task in BOINC requires either modifying its code to include calls to the BOINC API or using an existing wrapper. BOINC API contains the methods that must be executed by the tasks to start and finish execution, communicate progress, access input and output files, and report errors. The API's use is necessary only when there is no wrapper to run the application on a particular platform. The wrapper runs the applications as sub-processes and handles all communications between the BOINC client and the server.

When we started this research project, there was no compiled wrapper for Android OS. BOINC provides a generic wrapper that must be compiled for the target platform. Initially, there were wrappers for Windows, Linux, and Apple OS. The work presented in [48] explains the steps to compile the generic wrapper for ARM architecture with Android OS as well as other problems faced at execution time. The next paragraphs summarize the most relevant problems.

Cross-Compilation *Phones with the Android OS run on ARM CPUs and, to date, did not have a C++ native compiler pre-installed. Therefore, the generation of an executable code for the ARM architecture had to be done on a platform where* a C++ compiler is available, using a cross-compilation process. Although the BOINC tutorial contains information about the cross-compilation of applications, it had little documentation about compiling the generic wrapper for Android. Docker container helped us do the cross-compilation process since it has pre-built and configured toolchains for cross-compiling different platforms, including Android ARM (https://github.com/dockcross/dockcross).

Runtime Errors *Once the wrapper binary code is generated, its execution in Android versions equal to or greater than 5.0 generated a runtime error because the binary was not in PIE* (i.e., position-independent executable) format. These Android versions required this security feature (https://source.android.com/security/enhancements/enhancements50). The problem was solved by adding the flags *--fPIE -pie* to the compilation script.

The last runtime error was related to the PWD variable used in the wrapper code to get the current directory path. On Android, this variable does not exist or is not visible to the programmer. The solution was to modify the wrapper source code by changing the reference to PWD by the C function *getcwd*. With these new modifications, the wrapper worked properly. The compiled wrapper was placed in the official repository of BOINC source code (https://github.com/BOINC/boinc/pull/1671).

5.2   Generation of the Binary Code with the ITK Library

The ITK library has cross-platform support using the CMake build environment to manage the configuration process. ITK currently consists of more than 100 internal and other remote modules grouped in Core, ThirdParty, Filtering, IO, Bridge, Registration, Segmentation, Video, Compatibility, Remote External, and Numerics.

The generation of the binary code for the image processing requires the compilation and linking of functions from the ITK library. Since the code is in C++, the use of cross-compilation is also required. However, some ITK modules' cross-compilation is complex because it requires the execution of some binary code generated in the process; this intermediate execution is not possible because all the code is being generated inside the cross-compilation platform (e.g., Linux, Windows, etc.) and not in Android. This problem prevented us from cross-compiling the entire toolkit. The modules initially built were Core, Filtering, and Smoothing Recursive Gaussian Filter because they do not present the problem described above.

Also, during the cross-compilation process, it was necessary to deactivate the test flag BUILD_TESTING, to avoid the execution of some tests that are carried out at the end of the library construction to check the product. These tests assume that the system is running on the target platform, which is not true in this case.

The last problem was to produce a static executable code to replace the dynamic load of libraries. By default, the generated code loads libraries into its object factory and links against shared libraries. We chose to generate a self-contained static executable code since it is impossible to ensure that the target mobile devices have the necessary software installed to run the algorithms.

5.3   Parallelization of Image Processing

The parallelization model that BOINC currently supports is SPMD (single program multiple data), i.e., the same algorithm is executed on different pieces of the image. There is no communication between tasks, only between volunteers and the server. For the parallel processing, it was necessary to create programs to divide the image, automate the generation of workunits giving the corresponding parameters (i.e., parts of the image), collect the outputs, and generate the final result. The BOINC Work Generator and the Assimilator were extended to address these issues.

5.4   Performance Results

In [48], we described the experiments to evaluate the parallel solution's performance in mobile devices. In these experiments, we ran the Smoothing Recursive Gaussian Image Filter of ITK. This filter generates a smooth image by performing a convolution on it, which uses a kernel with values given by the Gaussian distribution. The filter was applied to a tiff image of 301 MB, 30576 pixels wide, and 9860 pixels high.

First, we executed a base test running the sequential algorithm in an 8 core Intel Core i7-920XM 20 2.0 GHz processor, 4GB of RAM, and Linux x86_64 Ubuntu 14.04 operating system. As indicated in Table 3, the running time of the base test was 49 min.

Then a 2^k experimental design allowed us to evaluate the effect of several factors on the performance. Factors selected were (a) execution platforms (mobile devices vs. desktop computers), (b) degree of parallelism (450 and 1000 workunits, each having image chunks of sizes 2 MB and 900 KB, respectively), (c) redundancy (1 or

Table 1 Portion of variation explained by the four factors and interactions

A: Execution platform	B: Degree of parallelism	C: Redundancy	D: Location of the server	AB	AC	AD	BC	BD	CD
0%	6%	47%	30%	2%	1%	1%	1%	0%	6%

Table 2 Execution times in minutes

		Desktops		Mobile devices	
		450 workunits	1000 workunits	450 workunits	1000 workunits
Local server	1 replica	15 min	37 min	19 min	42 min
	2 replicas	38 min	1 h 22 min	40 min	1 h 26 min
Remote server (cloud AWS)	1 replica	48 min	59 min	54 min	1 h 03 min
	2 replicas	1 h 37 min	1 h 54 min	1 h 52 min	2 h 01 min

Table 3 Sequential execution time and effect of the location of the server with the rest of the factors at their lowest level

	Linux (in a desktop computer)	Mobile devices – cloud server	Desktops – cloud server	Mobile devices – local server	Desktop computer – local server
Time (minutes)	49	54	48	19	15

2 replicas), and (d) server location (local or in the cloud). The response variable was the execution time.

Table 1 presents the percentage of variation explained by the four factors and interactions. According to the table, the factors that affect the most response variable are the server location and redundancy. Table 2 shows the execution times for factors and level combinations. Table 3 highlights the sequential execution time and the effect of the server's location with the rest of the factors at their lowest level.

The main conclusions of this study were:

The preliminary results demonstrated that the parallelism could be adequate but in lower degrees (compare results for 450 and 1000 workunits in Table 1). A more detailed performance evaluation of the parallel solution is left for future work.

When comparing execution times in desktop computers and mobile devices, we noticed no significant difference. The highest difference when BOINC is running at a local server was 26% (Table 1). The percentage of variation of this factor on the response variable is 0%. The best option may be to use both types of devices, when possible.

Although BOINC implements redundancy mainly for preventing errors in results, this technique has been proposed in scheduling algorithms for mobile grids to guarantee task completion in the context of disconnections, insufficient battery, etc. Our results suggested that the number of replications has a significant impact on execution times because it increases the number of workunits that must be sent and processed in the devices. These results were considered when we evaluated new BOINC scheduling strategies (see Sect. 6).

Finally, by looking at the application performance at different locations on the BOINC server, we noticed that the parallel solution's execution times with a local server were reduced by more than 50%, improving even the sequential solution (see Tables 2 and 3). When the performance is important, the recommendation is that both the server and the clients reside on the same local network. Although the subject should be studied in greater depth, this result favors local processing instead of uploading the code in the cloud, a solution proposed by other authors.

The next step in the research was to modify the BOINC scheduling strategy.

6 Proper Use of Devices

Many factors make the scheduling of tasks on mobile devices challenging. Some of these factors came from their characteristics: The battery power is limited, and wireless networks make communication feasible even while mobile devices are moving. However, wireless communications may be partially or totally interrupted at any time. Additionally, mobile devices are used by people who move and can disconnect, affecting the total execution times or the completion of tasks. These people own the resource and use it for other purposes; therefore, the devices cannot be considered dedicated platforms. Hence, running applications should make reasonable use of them.

Although BOINC allows the execution of tasks in mobile devices, some of the previously mentioned challenges are not fully resolved yet. For example, the devices' battery status is not considered in task scheduling, and replication is mainly used to avoid result errors. This led us to think about new scheduling strategies for BOINC that assure the completion of tasks despite the lack of battery or disconnections and making fair use of the devices' energy.

In the literature, there are many scheduling strategies for mobile devices. [49] cite several works and classify these strategies into two groups: preventive and reactive. Preventive scheduling tries to assure the task culmination either by selecting the most reliable resource for execution or by using redundant resources (replication). Some examples of preventive methods are presented in [29, 50–53]. On the other hand, reactive scheduling strategies act after the tasks have been assigned to the resources. There are at least two approaches: (1) reallocation of tasks when resources fail or are about to fail (e.g., [54, 55]) and (2) reallocation of tasks to do load balancing among different nodes and thus maximize the number of tasks that can be completed (e.g., [56]).

The authors of [57] present the concept of "Smart Mobile Devices (SMD) singularities" as the aspects that contribute to resource heterogeneity and dynamic availability. Based on singularities, the authors also propose a classification of scheduling strategies

This varied offering of algorithms was evaluated to select the most suitable option for BOINC and our requirements. In the selection, the following aspects were considered:

- *Operating system services*: The mobile client of BOINC runs on Android OS, which provides its own set of system calls, like any operating system. Any selected strategy should not depend on services not provided by Android OS.
- *Compatibility with the BOINC model*: Some characteristics of the BOINC model may affect the selection of the strategy. These characteristics are the following: (a) BOINC can be classified as a proxy mobile grid, where the executing nodes or volunteers communicate with a server that acts as a proxy by dispatching the tasks to be executed. (b) In the model of volunteer computing, the server does not autonomously assign the tasks to the clients. On the contrary, tasks are sent to computing nodes (mobile or fixed) only when they request them. (c) The BOINC server does not provide information to the executing nodes about other computing nodes or volunteers connected to the grid, even if they are executing workunits of the same application. (d) Mobile devices are not for the exclusive use of BOINC. Characteristic "a," for example, would prevent choosing a hierarchical strategy like the one described in [58]. Characteristic "d" would prevent the use of the approach presented in [59]. This algorithm modifies the processor's speed in an embedded system by applying a higher or lower voltage. In Android, it is possible to apply this type of modification so that the processor can be instructed to work at a higher or lower speed. However, this type of technique can affect the performance of some applications of daily use in smartphones, such as games, mail, etc.
- *Low overhead*: One of this project's objectives is to guarantee that the works in progress finish at the appropriate time. On the other hand, in a voluntary computing model, the computing nodes (fixed or mobile) request the jobs and expect the assignment to be done in the shortest possible time. For this reason, efficiency should be privileged in the selection of the strategy. Thus, a strategy that requires analyzing much historical data before making the assignment will not be suitable for our project.
- *Information about the strategy*: Having a greater detail of the algorithm will facilitate its implementation and avoid introducing errors. Also, the authors must clearly explain how the algorithm was validated.

Considering these criteria, the strategy that was evaluated with the highest score was the algorithm presented in [52]. This algorithm is part of the group of preventive strategies and considers the battery state for the assignment of tasks. We also selected a replica-based strategy to guarantee the completion of tasks (see [60]). Since replicas' use affects the performance of the application, it should be left for scenarios of high instability or intermittence.

The next paragraphs describe the implemented algorithm and preliminary performance results. A detailed description of the methodology to select new scheduling strategies, their implementation, and results can be found in [60]. We called BOINC-MGE (Mobile Grid Extension) the version of BOINC with the new strategies.

6.1 SEAS – Simple Energy-Aware Scheduler

In [52] a simple battery estimation model is presented. The proposed model's characteristics are as follows: (a) It does not require to know several battery physical parameters that are only known by manufacturers. (b) It is not based on complex mathematical models (e.g., differential equations, thus affecting the overall system performance). (c) Uses real-time collected information about battery consumption, which means that it can predict battery discharge rate even if the mobile device is not used exclusively to execute the tasks that are sent by the BOINC server.

The proposed algorithm obtains the current battery charge (bc) and the current time (ct), and then waits for a battery status change. When a change happens, the algorithm measures the new battery charge (nbc) and the new current time (nct), and with this information calculates the current discharge rate (dr) as shown in Eq. 1.

$$dr = \frac{bc - nbc}{nct - nc} \tag{1}$$

Equation 2 shows how rt is computed. This is the remaining time available for computing before battery exhaustion, assuming that the discharge rate is constant.

$$rt = \frac{nbc}{dr} \tag{2}$$

However, because the conditions of the use of a device change over time [61], the discharge rate is not always constant. Therefore, SEAS proposes to keep a record of the previous estimations (rt) and use a historical average to determine the available time of the device. Based on these previous estimations as well as the execution time of previous tasks on the device, BOINC can determine the number of tasks to be sent so that they can be completed before the device is disconnected due to a lack of battery.

6.2 Integrating the Scheduling Strategy into BOINC

Figure 1 shows the new task scheduling process. The send_work_host routine was modified to incorporate the SEAS strategy. This routine is always called when the BOINC-MGE extension is enabled in a volunteer project. The calc_num_replicas routine, on the other hand, is called only if replication is enabled. As was previously described, after selecting the tasks to be sent, the add_results_to_reply routine adds the tasks into the assigned_tasks array.

The Android client application (BOINC-MGE APP in Fig. 1) was also updated to add the ability to collect some metrics related to the battery level status and send them to the BOINC server using the request_tasks routine payload message. These metrics feed the SEAS strategy.

Volunteers

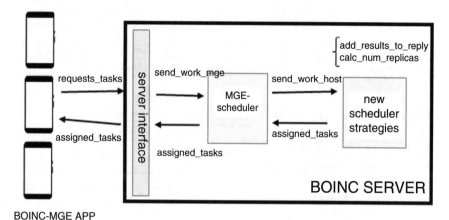

BOINC-MGE APP

Fig. 1 BOINC-MGE task scheduling process

Results To test the new scheduler strategies, we experimented with the following characteristics:

We write a benchmark composed of the canny_edge_app plus a component that forces a high CPU usage within the devices while executing the task. The app uses the well-known Canny edge algorithm [62] that segments images by extracting their edges.

The application processed large image files of more than 500 MB. Those files were divided into shorter images used later as the input files for the tasks generated and sent to the mobile devices. A total of 50 tasks are sent to the devices for processing.

The BOINC server was installed in a virtual machine of Google Cloud deployed in the United States. The virtual machine had Linux Debian as the operating system, 1 vCPU, 1.8GB of RAM, and 20GB of disk space. The mobile devices acting as volunteers had the next characteristics: 4 Samsung Galaxy S4, ARM Octa-Core processor at 1.6GHz, and a 2600 mAh battery and 1 Motorola Moto G, ARM Quad-Core processor at 1.2 GHz, and a 2070 mAh battery.

The $2^k r$ experimental design had the following components:

Response variables: the total execution time (i.e., the total time spent since the scheduler sent the first sub-image until the last segment's reception so the entire image can be reconstructed) and the average battery consumed by the devices after the completion of tasks.

Factors: the amount of battery in the volunteers and the use of the new strategy. Table 4 shows the factors and levels. The number of replicas of each experiment was 3.

Table 4 Factors and levels of the experimental design

Factors	Low level	High level
Version of BOINC	BOINC	BOINC-MGE (with the new scheduling strategy)
Status of the battery in the volunteers	Battery restrictions on some devices (2 of them have 20% of battery)	No battery restrictions on mobile devices

Table 5 Average battery consumption (in percentage) for every factor combination

	Average battery consumption
Limited battery level and BOINC	25
Limited battery level and BOINC-MGE	20
No battery limitation and BOINC	26
No battery limitation and BOINC-MGE	18.67

Table 6 Execution times in minutes for every factor combination

	Average execution time
Limited battery level and BOINC	97.56
Limited battery level and BOINC-MGE	66.69
No battery limitation and BOINC	77.59
No battery limitation and BOINC-MGE	52.58

Tables 5 and 6 show the average response variable values for each combination of factors and levels. Table 5 shows that BOINC-MGE (i.e., BOINC with the SEAS strategy) generates less battery consumption than its counterpart BOINC. The reason is that BOINC-MGE predicts which devices will end up disconnected in a battery-limited scenario. Based on this prediction, the scheduler avoids sending them tasks that will only be partially executed, resulting in an impact on the total battery consumed throughout the grid.

Table 6 presents the execution times. BOINC-MGE reduces total processing time by about 32% compared to BOINC. When there are battery limitations, the devices do not finish their tasks, and BOINC only finds out about this when a timeout expires. The timeout in this experiment is set to 30 min. Since BOINC-MGE predicts disconnections due to lack of battery, reassignment of tasks is not necessary. This saves waiting times in the server that affect the total execution times. When there is no battery limit, our solution is better because tasks are assigned and dispatched in advance. While BOINC does not dispatch tasks while others are running (occupying the device's CPUs), BOINC-MGE predicts that the mobile device can process more tasks and allocates them in advance. This saves time in downloading the necessary files for task processing since the download is done in parallel with other tasks' execution.

7 Conclusions and Future Work

This paper has presented challenges related to mobile devices' use in the distributed processing of medical images. We presented three challenges: (1) the choice of an execution platform that will meet the requirements of the project, (2) the transparency for the programmer, and (3) the proper use of devices for the execution of tasks, considering the intermittency, battery limitations, and the shared use, among others. The conclusions and learned lessons from our experience are as follows:

- There are many systems and algorithms in the literature that allow you to perform tasks on mobile devices. Although the number of articles is much greater than the available proposals, it is possible to find open-source systems that have been used in other types of projects and have a supportive community; this is the case of BOINC. The grid BOINC allowed us to get to know the technology faster, but we still wonder if its model is the most suitable for the project requirements.
- The research carried out so far leaves us with a set of important attributes to choose solutions (e.g., scheduling strategies) within the state of the art.
- BOINC could not be used directly, and we had to make several modifications to execute the algorithm for image parallel processing and select resources according to their battery availability. It is important to use scheduling strategies that consider the characteristics of mobile devices. This avoids re-scheduling, and therefore it is gained in time of execution and suitable use of the battery. Additionally, it is required to use strategies that guarantee the completion of tasks in contexts of high intermittency.
- Tools like Docker (https://www.docker.com/) support the task of developing programs on mobile devices.
- The performance obtained in the execution of mobile devices is promising, as well as the use of nearby devices for parallel processing instead of exchanging information with the cloud.
- The use of replicas does not seem recommended when we want efficiency in parallel processing, especially if any component is outside the local area network. However, in contexts of much intermittence, the replicas can guarantee that the tasks are completed.
- Although mobile devices currently have the required features to perform processing, their shared use with the owners affects their role in these types of projects in several ways: (a) It is expected that external tasks do not (heavily) degrade the performance of owners' applications or experience. Careful attention to this aspect could influence factors such as the scheduling algorithm, the parallelization strategy, the type and number of tasks sent, etc. (b) In the volunteer model, owners decide when to receive tasks. Depending on the use cases, it could be necessary to evaluate more invasive execution models. For example, the scheduler can select the devices belonging to a health institution at any time. (c) As much as possible, the software to be executed in mobile devices should be self-

contained. Dynamic library loads should be avoided as this affects performance and can be more difficult in intermittent scenarios.

- Although the BOINC model has served us to learn details about processing on mobile, it is important to continue studying in depth the cases of use of this technology and the possible options that the technology offers us: voluntary computing, CWC (computing while charging), mobile cloud computing, ad-hoc grids, etc.

Future work includes:

- Continue to conduct BOINC experiments that involve several types of clients (fixed and mobiles), different network configurations, and new scheduling strategies.
- Test new image processing algorithms and parallelization options. Finding the best parameters for parallel execution.
- Find the use cases for this technology.
- Based on the use cases, it is necessary to ask us again whether BOINC is still the most suitable technology. If not, look for other systems using the selection attributes useful in this first phase of the project.
- When adopting the voluntary computing model, it would be important to answer: Who would be the volunteers? What could be the incentives for them?

References

1. E. Agu, P. Pedersen, D. Strong, et al., The smartphone as a medical device: Assessing enablers, benefits and challenges, in *IEEE International Workshop of Internet-of-Things Networking and Control, IoT-NC 2013*, (2013). https://doi.org/10.1109/IoT-NC.2013.6694053
2. D.C. Baumgart, Smartphones in clinical practice, medical education, and research. Arch. Intern. Med. **171**(14), 1294–1296 (2011)
3. M.G. Masciantonio, A.A. Surmanski, Medical smartphone applications A new and innovative way to manage health conditions from the palm of your hand. Univ. West. Ont. Med. J. **86**(2), 51–53 (2017)
4. J.C. Contreras-Naranjo, Q. Wei, A. Ozcan, Mobile phone-based microscopy, sensing, and diagnostics. IEEE J. Sel. Top. Quantum Electron. **22**(3), 1–14 (2015)
5. R. Rajendran, J. Rajendiran, Image analysis using smartphones for medical applications: A survey. Intell. Pervasive Comput. Syst. Smarter Healthcare, 275–290 (2019)
6. T.M. Mengistu, D. Che, Survey and taxonomy of volunteer computing. ACM Comput. Surv. **52**(3), 1–35 (2019). https://doi.org/10.1145/3320073
7. J.M. Rodríguez, C. Mateos, A. Zunino, Are smartphones really useful for scientific computing? in *International Conference on Advances in New Technologies, Interactive Interfaces, and Communicability*, (Springer, Berlin, Heidelberg, 2011), pp. 38–47
8. L. Duan, T. Kubo, K. Sugiyama, J. Huang, et al., Motivating smartphone collaboration in data acquisition and distributed computing. IEEE Trans. Mob. Comput. **13**(10), 2320–2333 (2014). https://doi.org/10.1109/TMC.2014.2307327
9. M.Y. Arslan, I. Singh, S. Singh, H.V. Madhyastha, et al., CWC: A distributed computing infrastructure using smartphones. IEEE Trans. Mob. Comput. **14**(8), 1587–1600 (2015). https://doi.org/10.1109/TMC.2014.2362753

10. H. Kurdi, M. Li, H. Al-Raweshidy, A classification of emerging and traditional grid systems. IEEE Distrib. Syst. Online **9**(3), 1 (2008)
11. J. Furthmüller, O.P. Waldhorst, Survey on Grid computing on mobile consumer devices, in *Handbook of Research on P2P and Grid Systems for Service-Oriented Computing: Models, Methodologies and Applications*, (IGI Global, 2010), pp. 313–337
12. M. Hijab, D. Avula, Resource discovery in wireless, mobile and ad hoc grids – Issues and challenges, in *International Conference on Advanced Communication Technology, ICACT*, (2011), pp. 502–505
13. I. Foster, C. Kesselman, S. Tuecke, The anatomy of the grid: Enabling scalable virtual organizations. Int. J. High Perform. Comput. Appl. **15**(3), 200–222 (2001)
14. D.C. Marinescu, G.M. Marinescu, Y. Ji, et al., Ad hoc grids: Communication and computing in a power constrained environment, in *IEEE International Performance, Computing and Communications Conference*, (2003), pp. 113–122. https://doi.org/10.1109/pccc.2003.1203690
15. L.F.G Sarmenta, Volunteer computing, Doctoral Dissertation, Massachusetts Institute of Technology, 2001
16. H.T. Dinh, C. Lee, D. Niyato, P. Wang, A survey of mobile cloud computing: Architecture, applications, and approaches. Wirel. Commun. Mob. Comput. **13**(18), 1587–1611 (2013)
17. N. Fernando, S.W. Loke, W. Rahayu, Mobile cloud computing: A survey. Futur. Gener. Comput. Syst. **29**(1), 84–106 (2013)
18. C. Mateos, A. Zunino, M. Campo, A survey on approaches to gridification. Software Pract. Exp. **38**(5), 523–556 (2008)
19. K. Källander, J.K. Tibenderana, O.J. Akpogheneta, Strachan, et al., Mobile health (mHealth) approaches and lessons for increased performance and retention of community health workers in low-and middle-income countries: A review. J. Med. Internet Res. **15**(1), e17 (2013)
20. M. Has, A.B. Kaplan, B. Dizdaroğlu, Medical image segmentation with active contour model: Smartphone application based on client-server communication, in *2015 Medical Technologies National Conference (TIPTEKNO)*, (2015), pp. 1–4
21. T. Kitrungrotsakul, C. Dong, T. Tateyama, Han, et al., Interactive segmentation and visualization system for medical images on mobile devices. J. Adv. Simul. Sci. Eng. **2**(1), 96–107 (2015)
22. P. Vaish, R. Bharath, P. Rajalakshmi, U.B. Desai, Smartphone-based automatic abnormality detection of kidney in ultrasound images, in *IEEE 18th International Conference on e-Health Networking, Applications and Services (Healthcom)*, (2016)
23. C. Luo, Patient-centered monitoring and image processing on smartphones, Doctoral Dissertation, Texas A&M University, 2017
24. Y. Chen, Q. Zhao, X. Hu, B. Hu, Multi-resolution parallel magnetic resonance image reconstruction in mobile computing-based IoT. IEEE Access **7**, 15623–15633 (2019)
25. Y. Zhuang, N. Jiang, Z. Wu, Q. Li, et al., Efficient and robust large medical image retrieval in mobile cloud computing environment. Inf. Sci. **263**, 60–86 (2014)
26. D. Vazhenin, Cloud-based web-service for health 2.0, in *Proceedings of the 2012 Joint International Conference on Human-Centered Computer Environments (HCCE'12), New York*, (2012)
27. L. Liu, W. Chen, M. Nie, Zhang, et al., iMAGE cloud: medical image processing as a service for regional healthcare in a hybrid cloud environment. Environ. Health Prev. Med. **21**(6), 563–571 (2016)
28. F. Büsching, S. Schildt, L. Wolf, Droidcluster: Towards smartphone cluster computing–the streets are paved with potential computer clusters, in *2012 32nd International Conference on Distributed Computing Systems Workshops*, (2012), pp. 114–117
29. P. Datta, S. Dey, H.S. Paul, A. Mukherjee, ANGELS: A framework for mobile grids, in *2014 Applications and Innovations in Mobile Computing (AIMoC)*, (2014), pp. 15–20
30. C. Loos, E-health with mobile grids: The Akogrimo heart monitoring and emergency scenario, Akogrimo White Paper (Online), 2006
31. A. Coronato, G. De Pietro, MiPeG: A middleware infrastructure for pervasive grids. Futur. Gener. Comput. Syst. **24**(1), 17–29 (2008)

32. A. Wolff, S. Michaelis, J. Schmutzler, C. Wietfeld, Network-centric middleware for service oriented architectures across heterogeneous embedded systems, in *2007 Eleventh International IEEE EDOC Conference Workshop*, (2007), pp. 105–108
33. A. Litke, D. Halkos, K. Tserpes, D. Kyriazis, Fault tolerant and prioritized scheduling in OGSA-based mobile Grids. Concurrency Comput. Pract. Exp. **21**(4), 533–556 (2009)
34. M. Alia, F. Eliassen, S. Hallsteinsen, E. Stav, Madam: towards a flexible planning-based middleware, in *Proceedings of the 2006 International Workshop on Self-Adaptation and self-Managing Systems*, (2006), p. 96
35. D. Anderson, BOINC: A system for public-resource computing and storage, in *Proc. IEEE/ACM Int. Work. Grid Computing*, (2004)
36. N. Palmer, R. Kemp, T. Kielmann, H. Bal, Ibis for mobility: Solving challenges of mobile computing using grid techniques, in *Proceedings of the 10th Workshop on MOBILE Computing Systems and Applications*, (2009)
37. S. dos Lima, L. Gomes, A.T. Ziviani, A. Endler, et al., Peer-to-peer resource discovery in mobile grids, in *Proceedings of the 3rd International Workshop on Middleware for Grid Computing*, (2005)
38. N. Andrade, W. Cirne, F. Brasileiro, P. Roisenberg, OurGrid: An approach to easily assemble grids with equitable resource sharing, in *Workshop on Job Scheduling Strategies for Parallel Processing*, (Springer, Berlin, Heidelberg, 2003), pp. 61–86
39. R. García, L. Florez-Valencia, M. Curiel, On existing mobile grids for android devices, in *2016 8th Euro American Conference on Telematics and Information Systems (EATIS)*, (2016), pp. 1–7
40. S. Schildt, F. Büsching, E. Jörns, L. Wolf, CANDIS: Heterogenous mobile cloud framework and energy cost-aware scheduling, in *In: Proceedings – EEE International Conference on Green Computing and Communications and IEEE Internet of Things and IEEE Cyber, Physical and Social Computing, GreenCom-IThings-CPSCom*, (2013), pp. 1986–1991
41. J. Wagner, X. Cao, A prototype for distributed computing platform using smartphones, in *Proceedings of the Conference on Information Systems Applied Research*, (2020) ISSN, 2167, 1508.
42. M.P. Kumar, R.R. Bhat, S.R. Alavandar, V.S. Ananthanarayana, Distributed public computing and storage using mobile devices, in *2018 IEEE Distributed Computing, VLSI, Electrical Circuits and Robotics (DISCOVER)*, (2018), pp. 82–87
43. E. Cuervo, P. Gilbert, B. Wu, L.P. Cox, CrowdLab: An architecture for volunteer mobile testbeds, in *2011 Third International Conference on Communication Systems and Networks (COMSNETS 2011)*, (2011), pp. 1–10
44. H. Ba, W. Heinzelman, C.A. Janssen, J. Shi, Mobile computing-A green computing resource, in *2013 IEEE Wireless Communications and Networking Conference (WCNC)*, (2013), pp. 4451–4456
45. K. Habak, M. Ammar, K.A. Harras, E. Zegura, Femto clouds: Leveraging mobile devices to provide cloud service at the edge, in *2015 IEEE 8th International Conference on Cloud Computing*, (2015, June), pp. 9–16
46. D.P. Anderson, K. Reed, Celebrating diversity in volunteer computing, in *2009 42nd Hawaii International Conference on System Sciences*, (2009), pp. 1–8
47. M. Taufer, D. Anderson, P. Cicotti, C.L. Brooks, Homogeneous redundancy: A technique to ensure integrity of molecular simulation results using public computing, in *19th IEEE International Parallel and Distributed Processing Symposium*, (2005)
48. M. Curiel, D.F. Calle, A.S. Santamaría, Suarez, et al., Parallel processing of images in mobile devices using BOINC. Open Eng. **8**(1), 87–101 (2018)
49. M. Curiel, Wireless grids: Recent advances in resource and job management, in *Handbook of Research on Next Generation Mobile Communication Systems*, (IGI Global, 2016), pp. 293–320
50. C. Li, L. Li, Utility-based scheduling for grid computing under constraints of energy budget and deadline. Comp. Stand. Interfaces **31**(6), 1131–1142 (2009)

51. S.S. Vaithiya, S.M.S. Bhanu, Mobility and battery power prediction based job scheduling in mobile grid environment, in *International Conference on Parallel Distributed Computing Technologies and Applications*, (Springer, Berlin, Heidelberg, 2011), pp. 312–322
52. J.M. Rodriguez, A. Zunino, M. Campo, Mobile grid seas: Simple energy-aware scheduler, in *Proc. 3rd High-Performance Computing Symposium-39th JAIIO*, (2010)
53. P. Ghosh, S.K. Das, Mobility-aware cost-efficient job scheduling for single-class grid jobs in a generic mobile grid architecture. Futur. Gener. Comput. Syst. **26**(8), 1356–1367 (2010)
54. D. Lee, S. Chin, J. Gil, Efficient resource management and task migration in mobile grid environments, in *International Conference on Security-Enriched Urban Computing and Smart Grid*, (Springer, Berlin, Heidelberg, 2010), pp. 384–393
55. A. Adeyelu, E. Olajubu, A. Aderounmu, T. Ge, A model for coordinating jobs on mobile wireless computational grids. Int. J. Comput. Appl. **84**(13), 17–24 (2013)
56. J.M. Rodriguez, C. Mateos, A. Zunino, Energy-efficient job stealing for CPU-intensive processing in mobile devices. Computing **96**(2), 87–117 (2014)
57. M. Hirsch, C. Mateos, A. Zunino, Augmenting computing capabilities at the edge by jointly exploiting mobile devices: A survey. Futur. Gener. Comput. Syst. **88**, 644–662 (2018)
58. C.Q. Huang, Z.T. Zhu, Y.H. Wu, Z.H. Xiao, Power-aware hierarchical scheduling with respect to resource intermittence in wireless grids, in *2006 International Conference on Machine Learning and Cybernetics*, (2006), pp. 693–698
59. T.A. AlEnawy, H. Aydin, Energy-constrained scheduling for weakly-hard real-time systems, in *26th IEEE International Real-Time Systems Symposium (RTSS'05)*, (2005)
60. A.S. García Payares, BOINC-MGE mobile grid extension, Master Thesis, Pontificia Universidad Javeriana, Bogotá, 2019
61. W.X. Shen, C.C. Chan, E.W.C. Lo, K.T. Chau, Estimation of battery available capacity under variable discharge currents. J. Power Sources **103**(2), 180–187 (2002)
62. J. Canny, A computational approach to edge detection. IEEE Trans. Pattern Anal. Mach. Intell. **6**, 679–698 (1986)

Smart Traffic Control for Emergency Vehicles Using the Internet of Things and Image Processing

Sandesh Kumar Srivastava, Anshul Singh, Ruqaiya Khanam, Prashant Johri, Arya Siddhartha Gupta, and Gaurav Kumar

1 Introduction

As uncovered by the National Institute of Emergency Medicine (NIEM), approximately 20% of patients requiring crisis treatment have passed away on route to the clinic due to delays because of gridlock and uncooperative drivers. To reduce the concerning proportion of such deaths, some countries have opted for a separate lane system for emergency vehicles, but not all countries can benefit from the same solution. Some countries are sufficiently populated that despite having certain facilities, emergency vehicles cannot reach the desired location on time. This chapter proposes a model that addresses all such problems. The most fitting answer for the blockage issue is letting the clogged sides cross the convergence first. The thought incorporates finding the ideal path for the emergency vehicle from source to objective in the ongoing rush hour gridlock issue utilizing Dijkstra's algorithm on the weighted diagram of continuous traffic information. The model further uses the internet of things and digital image processing to make it work more conveniently [1].

1.1 Image Processing

Digital image processing examines and deciphers captured or recorded information. Picture advancement, information mining, and feature extraction are the three significant components of digital image processing.

S. K. Srivastava (✉) · A. Singh · P. Johri · A. S. Gupta · G. Kumar
Galgotias University, Greater Noida, India

R. Khanam
Sharda University, Greater Noida, India

© Springer Nature Switzerland AG 2022
P. Johri et al. (eds.), *Trends and Advancements of Image Processing and its Applications*, EAI/Springer Innovations in Communication and Computing, https://doi.org/10.1007/978-3-030-75945-2_3

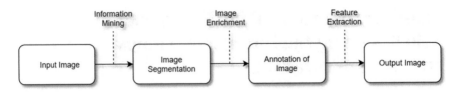

Fig. 1 Image processing flow diagram

The image enrichment procedure is utilized to improve the perceivability of the highlights, stifling unnecessary data. Mining is utilized to obtain the factual information of the detailed situation of the picture. The model comprises an intersection point between the camera and the sensor framework. The electrical or the optical sign from this sensor is changed over to the advanced sign. The signal invokes the capture of a picture in real-time and sends it to the dataset for processing. Then the image is matched, and if it matches any of the dataset entries, then the algorithm follows the path of taking the difference and hence counting the heads based on this difference image matching [2] (Fig. 1).

Use of Image Processing

The cameras, introduced on all the cross-ways, take ongoing pictures of the traffic. The noise, for example, Gaussian noise or salt and pepper noise, is eliminated.

The calculation to eliminate the Gaussian noise in computerized pictures is a basic and quick non-direct strategy. The presentation of the new channel is tried for different noise undermined grayscale and shading pictures. The rebuilding consequences of the pre-owned calculation are contrasted with standard strategies and recently proposed techniques. It eliminates Gaussian noise with edge conservation for low to high Gaussian noise tainted pictures. Test results show that the proposed calculation has less mean absolute error (MAE) and higher peak signal-to-noise proportion (PSNR) than the mean channel, wiener channel, alpha-managed channel, K implies calculation, reciprocal channel, and the recently proposed three-dimensional channel. The calculation has less computational time and multifaceted nature, and subsequently, it can likewise be executed in equipment [3].

Middle sifting is utilized for eliminating salt and pepper noise subsequent to eliminating Gaussian noise. This sifting is less delicate than direct procedures to extraordinary changes in pixel values; it can eliminate salt and pepper noise without essentially diminishing the sharpness of a picture [3].

This image further will be compared with the server-stored data to identify the density change, and hence the headcount of the vehicles in all lanes will be counted accordingly.

1.2   Internet of Things

The internet of things, commonly known as IoT, is a system of interrelated devices that can share the data between different systems without the help of human resources. Every device has its own unique id (UID); therefore, it is easy to control and manage a device by setting a place. In transportation, speed tracking is done using the IoT [4].

Use of the Internet of Things

Implementation of the internet of things could be done with the help of IoT chips. Every traffic signal will have several chips installed according to the number of cameras. The chip is used to count the number of vehicles, to track the emergency vehicle with the help of a GPS tracker, and to update the timer of traffic lights accordingly. This then counts the number of vehicles based on image processing and alters the waiting timer according to the traffic [5].

The framework would essentially zero in on the picture captured utilizing the camera. The captured picture would be cross-checked with a preset picture stored in the worker to recognize the density. In view of the density, the traffic developments are a trigger for the intersections. This decreases the general delay time and results in a smoother traffic stream. Consequently, the framework would work dependent on the assortment of density picture sent from the area to the worker [6].

Image processing techniques and the IoT are simultaneously used in a way that the paths with emergency vehicles will be given the higher preference and traffic will be managed accordingly. IoT chips installed in the traffic lights are used to alter the timer accordingly based on the weights given to regular vehicles and emergency vehicles.

IoT chips can be notified of the incoming of the emergency vehicle by connecting the emergency vehicle to the local server, as recognizing emergency vehicles from camera output may increase the complexity of the analysis. After connecting the emergency vehicle to the local server, the local server can then select the desired route to reach the destination in less time after analyzing the real-time traffic data on all the intersections. Both the emergency vehicle and IoT chip will be notified about the desired path and therefore the lane in the route of the emergency vehicle will be given more preference on all the intersections to enable it to reach the destination in less time.

2   Literature Review

The IoT has not only touched the various aspects of daily life but also revolutionized the way we work. The increase in the use of the IoT in areas of agriculture, health, and lifestyle such as security and creating a user-friendly environment has completely changed the phase of developments.

Dorothy et al. [7] demonstrate in their research on home security that image processing used together with the IoT can be used to develop a more secure environment. They also mention the use of the Cooley–Tukey algorithm, which helps compare images quickly and retrieve the results with greater accuracy.

The field of agriculture has made major advancements using the IoT because it offers an innovative solution to the problems faced in agriculture. For example, concerning watering the plants in fields or predicting rainfall or analyzing the fertility of the soil, the IoT has completely changed modern agriculture. As stated by Farooq et al. and Anugraheni et al., the work in agriculture is being patronized by governments of various countries, which makes it one of the quickest rising fields using the IoT [8, 9].

The growth of the IoT in traffic management and the use of sensors to assist traffic has contributed to making the roads a much safer place. Authors such as Frank et al. & Rane et al. have developed various algorithms that show us the combined use of the IoT and image processing in road safety, parking management, security, and various other fields [10, 11].

In [12] many IR sensors are inserted to calculate the number of vehicles per road route and by recording statistics in the cloud using Bluetooth connectivity, traffic bulk data is included in the compilation of derived algorithms in the KNN algorithm to obtain the required expected time for traffic lights. Using short-distance communication technology such as Bluetooth requires access points near the list of sensors to achieve data transfer and thus increases the complexity of the system. In addition, using an integration algorithm according to KNN leads to the overriding of the computer system, which can cause delays in decision-making and conversion time for robots, which will negatively be reflected in traffic.

Another proposition for a traffic signaling mark framework utilizing ultrasonic sensors was created in [13], where ultrasonic sensors occur each 50 meters of street to catch gridlock and contact Arduino to control traffic lights appropriately, and mass data is sent through Wi-Fi Raspberry Pi 3 where the investigation made on more traffic and less traffic than the day and time and the equivalent is sent to a web page that can be seen by traffic police authorities for additional examination.

To decrease the danger of gridlock in crises, particularly in ambulances [14], proposed an inserted emergency vehicle framework comprising Arduino Uno, GPS Arduino rescue vehicle following shield, and GSM Arduino shield to refresh the rescue vehicle area in the information base of the web. The framework tracks the path of the emergency vehicle and controls the traffic signals to guarantee that street intersections are free from traffic, making it simpler for ambulances to pass easily and immediately; however, this proposition does not consider the measure of traffic and time needed to arrive at the street signal. There was a mistake during the signal, which increased congestion of different points or postponements in departure circumstances.

Fahim and Sillito also suggested using the various anomaly detection and prediction technology in the IoT environment [15].

3   Proposed Work

The entire problem is defined under the subheadings that follow, and hence we divide the problem into smaller problems to compute the solution with greater efficiency.

3.1   Traffic Analyzation at Traffic Signal

The most optimal solution to the congestion problem is to clear out the more congested side across the intersection first. However, the congestion problem is not the only factor that needs to be considered for the timer setting of the lights. It is necessary to minimize the time duration for the journey of an emergency vehicle from source to its destination and to clear out the traffic very efficiently so that there is no congestion problem just because of the presence of emergency vehicles. This can be done by taking the headcount of the vehicles at the intersection parallelly by placing a camera on each side of the intersection and processing the images to obtain the number of vehicles on each side of the intersection [16].

There are two approaches to implement the above problem statement:

(a) *OpenCV approach*

The OpenCV is a library in python that helps in the process of image recognition. It works by converting the image to the desired size and computing the number of cars based on the percentage change among the different datasets given for the model. The camera is used to capture the traffic and the image taken is processed as follows in the real-time environment [17]:

- Examine using OpenCV library in python
- Resize as indicated by the need
- Convert the image to black and white
- Remove the noises
- Dilation
- Take distinction from reference images
- Discover the rate change
- Obtain the quantity of vehicles

(b) *Machine learning approach*

The machine learning module can be directly implemented in python using the inbuilt libraries such as Pandas and NumPy. The image taken by the camera is given to the model, which trains itself in real-time and undergoes the following operations:

- Preprocessed to decide its path
- Either a model is acquired or another one is made to recognize vehicles and various vehicles

- Picture is stacked and all the identifications are spared and later analyzed
- Vehicles are included in the identifications

Note

To make the system more dependable, the model should be able to predict the number of vehicles in all the light conditions accurately. With regard to how both of the approaches work, the prediction of the machine learning model is highly dependent on what type of dataset it is trained on, whereas the OpenCV model works by the processing of the images in real-time by converting the images into black and white and then reducing the noises, which makes it more reliable in non-ideal light conditions. Here, OpenCV seems to be the more reliable model after considering the odds [18].

The proposed system aims to give a comprehensive application to powerfully overseeing traffic signal systems based on current vehicle calculations. The Gaussian potential model is used to classify vehicles, while Kulman filters are used to recognize vehicles at the intersection of traffic signal timers. The proposed system takes all the roadways at the intersection as contribution from the computer's local storage and compares them based on the quantity of cars in the corresponding path. The highest density path is preferred. The specific path scenario is also considered when building up the system. It implements a density-based operation system using OpenCV video handling.

Image Acquisition The system starts with an image acquisition process in which live video is collected and processed by a camera set up on a signal stand.

Image Cropping Frames are taken from the video, which are additionally processed. The second step is image cropping, which focuses on the territory where vehicles are found and around noise and other information, helping to accomplish ROI for cropping methods, which helps in accomplishing high accuracy.

Background Reduction OpenCV implements three similar algorithms. For the proposed system, MOG background reduction will be used. It uses a strategy to demonstrate each background pixel from a combination of K Gaussian distributions ($K = 3$–5). The weight of the mixture indicates the time ratio in which those colors are in view. Possible background colors are high and consistent. When coding, we have to make a background object using a capacity called cv2.createBackground-SubtractorMOG (). It has some discretionary parameters such as length of history, Gaussian blend, and passage. It is set to certain default values. Inside the video circle, use the background subtractor.apply () technique to obtain the front mask (Fig. 2).

To get objects that are moving on the video:

foreground_objects = current_frame - background_layer

However, in some cases, we cannot obtain a static frame because the lighting may change, or some objects are moved by someone, or there is always development, and so on. In such cases, we save some numbers and attempt to discover which pixels are almost indistinguishable, and at that point, these pixels become

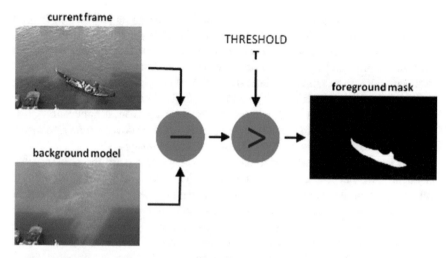

Fig. 2 Foreground object detection

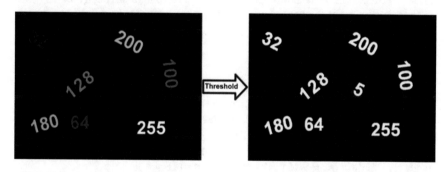

Fig. 3 Image thresholding

part of the background layer. Usually, the contrast between how we accomplish this background layer and the additional channel is that we use it to make the selection more precise [19].

Image Thresholding

Thresholding is used to classify pixel values as the image used on grayscale images, which are images with pixel values from 0 to 255. At the point when you insert an image, you classify the pixels that are set above and beneath each gathering. Thresholding should be possible through local methods as well as worldwide methods. Thresholding is one of the methods used to suppress the background and accomplish a reasonable front (Fig. 3).

Image thresholding is used to distance the image based on pixel intensity. The contribution of such an information calculation is usually a grayscale image and a threshold. The yield is a paired image. On the off chance that the intensity of the pixel in the information image exceeds the cutoff, the corresponding yield pixel is stamped white (front) and if the intensity of the info pixel is less than or equivalent to the information, the yield pixel position is dark (background) [20].

Image thresholding is used as a pre-processing step in most applications. For instance, you can use it in medical image processing to uncover a tumor on mammograms or to localize a cataclysmic event in satellite images.

One issue with general limitations is that you must specify the section value physically. We can physically check how great the passage is by attempting various values; however, it is laborious and it breaks down in reality. Therefore, we need an approach to set the section naturally. Otsu's innovation is named after its maker Nobu Otsu, and is a genuine case of auto thresholding.

In spite of the fact that the strategy was first reported in 1979, it remains the basis for some intricate solutions. The authors have given an improved object attribute thresholding (OAT) strategy as a means of estimating the localization of submerged land. The advantages of such a methodology contrasted with precise ongoing segmentation and threshold segmentation methods of submerged facilities are demonstrated in execution. To beat this spot challenge, a technique has been created based on the binarization of OATs, which limits the search scope to the suitable segmentation limit for the anterior object segmentation.

Method Implementation:
The image below is used as info (Fig. 4):

Read Image.

First, we read the image in grayscale and possible improvement with Gaussian blur to lessen the noise is determined. We have visualized the results of the preprocessed image and its histogram (Fig. 5):

In the histogram beneath, we can see an expressed mono peak and its close to locale and slightly expressed peak toward the start (Fig. 6):

Second, we have to compute the Otsu's threshold. To apply Otsu's method, we just need to use OpenCV threshold work with function THRESH_OTSU flag (Fig. 7).

The threshold esteem is close to that acquired above in a handmade case (131.98). Presently, we should see the last binarized image after Otsu's technique application (Fig. 8):

We can observe that the background and the principle objects in the image were separated. We have drawn a histogram for the acquired binarized image (Fig. 9).

As we can see, image pixels are presently separated in two clusters with intensity values 0 and 255 [21].

Filtering Morphological transformations are some normal operations relying upon the size of the image. This usually occurs on double images. It requires two inputs, one is our unique image, and the other is known as the structuring component or portion that determines the idea of the operation. The two basic syntax operators are

Fig. 4 Original image

Fig. 5 Grayscale image with Gaussian blur

erosion and weakening. At that point, its various forms of opening, closing, gradient, and so on also come into force [22].

Erosion The basic thought of erosion is similar to a soil cut, which removes the boundaries of the front article. The kernel slides through the image (as in 2D convolution). The pixel in the first image (1 or 0) is considered to be 1 if all the pixels underneath the kernel are 1; otherwise, it is erased (made zero). All pixels close to the border are discarded relying upon the kernel size. Therefore, the thickness or

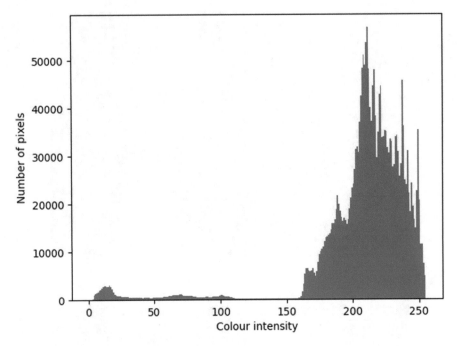

Fig. 6 Histogram of color intensity vs pixels for grayscale Gaussian blurred image in Fig. 5

```
1  ▦ Applying Otsu's method setting the flag value into cv.THRESH_OTSU.
2  # Use a bimodal image as an input.
3  # Optimal threshold value is determined automatically.
4  otsu_threshold, image_result = cv2.threshold(
5      image, 0, 255, cv2.THRESH_BINARY + cv2.THRESH_OTSU,
6  )
7  print("Obtained threshold: ", otsu_threshold)
```

Fig. 7 Pseudocode to compute Otsu's threshold

size of the foreground decreases or the white region in the image decreases. This is useful for eliminating small white noise, for separating two connected objects (Fig. 10).

Dilation It is the specific opposite of corrosion. Here, the pixel component is "1" if the pixel is in any event one pixel "1." Therefore, it increases the white territory of the image or increases the size of the foreground. Normally, in cases such as noise removal, scattering occurs subsequent to cutting. Because the cut eliminates white noise, it also reduces our item. Thus, we weaken it. Because of the noise, they do not return; however, our object area increases. It is also useful for interfacing broken parts of an object (Fig. 11).

Opening Opening is another name for erosion, which is followed by dilation. It is used in eliminating noise, as clarified previously (Fig. 12).

Fig. 8 Final object detected after Otsu's threshold calculation

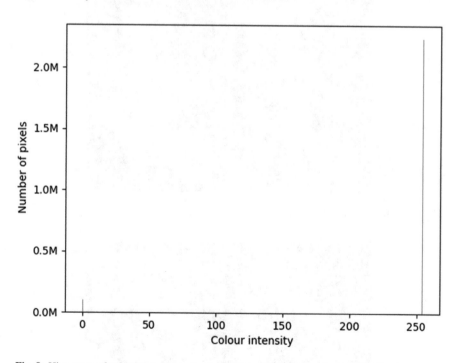

Fig. 9 Histogram of color intensity vs pixels after Otsu's threshold calculation in Fig. 8

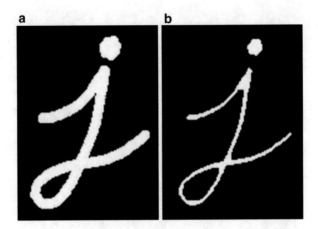

Fig. 10 (**a, b**) Image before and after erosion is done

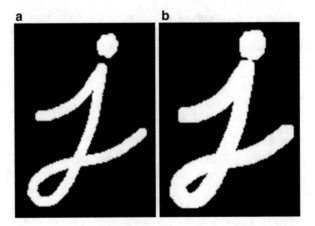

Fig. 11 (**a, b**) Image before and after dilation

Fig. 12 Image before and after opening

Fig. 13 Image before and after closing

Closing Closing is the reverse of opening—dilation before erosion. It is used in closing small holes inside the foreground image, or small dark dots on the image (Fig. 13).

Contour The binary image obtained from the throttling phase after filtration then comes through filtering to describe the texture for the identified object. Contrast could be described as a curve that meets each continuous point with the same shade or intensity. It helps for object recognition.

The find contours function in OpenCV helps us identify the contours. Similarly, the draw contours function helps us draw the contours.

Building Processing Pipeline (Fig. 14):
Class vehicles are used to store tracked vehicle information. The class vehicle count stores a list of traced vehicles and tracks the total vehicles seen. In each frame, the border-box and the position of the identified vehicles are used to apprise the status of the vehicle:

- Update currently tracked vehicles:

  For each vehicle:
  If there is a valid match for a given vehicle, update the condition of the vehicle and reset its final look counter. Remove the match from the candidate list
  Otherwise, increase the last-seen counter

- Create new vehicles for the leftover matches
- Update vehicle count

  For each vehicle:
  If the vehicle is a previous divider and not yet counted, update the total calculation and locate the numeric vehicle

Fig. 14 Workflow of image
processing

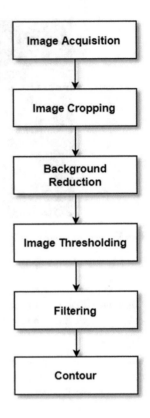

- Remove vehicles that are no longer in the frame

  For each vehicle:
  If the last-seen counter exceeds the threshold, remove the vehicle

  Furthermore, certain optimizations need to be done that will improve the quality
  of scanning the vehicle headcount number [23]. The factors that need to be considered for the optimization are:

  Cleaning up the foreground mask, tracking vehicles between the frames, and
  median blending (to improvise the quality of reading the objects).

  The selected model is to be implemented on each side of the intersection provided with the live feed and the data has to be retrieved from the IoT chip of each
  traffic signal. The IoT chips are to be set up on all sides of an intersection so that the
  traffic analyzation can be done on each side [24] (Fig. 15).

Fig. 15 Lane detection method

3.2 To Update a Load of Traffic to the Main Server

All the IoT chips are to be connected to the same Wi-Fi service provided by the city so that the real-time tracking of the traffic can be done. The number of vehicles counted by each camera on the intersection is to be updated by the IoT chip to the server and the actual weight calculation is to be maintained of each path on the map. After certain feedback is received from the chips, the congestion factor is to be maintained in the form of the directed graph data structure; a graph which is similar to the road network of the city. The nodes of the graph resemble the various intersections and the edges of the graph are the path of one intersection point to another. The weight of the edges is to be modified in real-time based on the feedback provided by each IoT chip [25] (Fig. 16).

3.3 Updating the Node Weights on the Map

Node weight is the value that represents the congestion factor of a single path. For a single path, there will be two sides: up and down, i.e., two intersections will be connected by a single path, but to keep the track of the traffic, this path will be

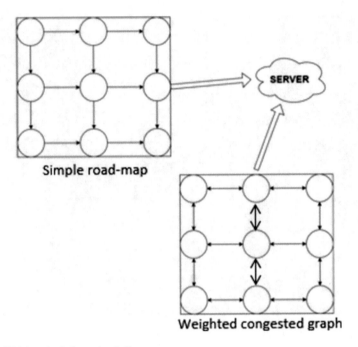

Fig. 16 Weight calculation using IoT

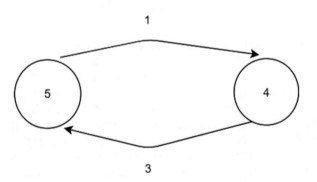

Fig. 17 Node updating

analyzed as per the up-going and down-going traffic. Hence, two junctions will be connected as described in Fig. 17.

To keep the congestion of traffic updated, every node is dependent on the feedback of a single IoT chip. The minimum possible node weight allowed is 0, as the number of vehicles cannot be negative. The dynamic directed graph is to be made available to the emergency vehicles such as ambulances, police cars, and fire brigades. This will allow emergency vehicles to retrieve the details about the real-time traffic in the city.

3.4 Update the Path Required to Be Followed by an Emergency Vehicle in Real-Time

The emergency vehicle's need is to reach its destination in the least time possible. This can be achieved by using Dijkstra's algorithm on the data fetched by the servers. By doing this, the least congested path can be achieved, but the need for the emergency vehicle is also to reduce the distance as much as possible. The least distance possible can be calculated by applying Dijkstra's algorithm to the equally weighted road-map graph of the city.

Let say in the given figure, if ABCD is the least congested path, but it is visible that CD is the more optimal solution because it joins the two intersections with the least distance possible, then the optimal solution can be calculated by combining the maximum number of common nodes obtained by applying Dijkstra's algorithm on both of the graphs giving more priority to the least distance solution [25] (Fig. 18).

3.5 Update the IoT Chip about the Emergency Vehicle

Once the path is finalized by the emergency vehicle, it has to be updated to the servers that one particular vehicle has to be given preference on one particular defined route. The chips are to be updated accordingly that an emergency vehicle will be passing through the route. The number can be more than one on a single path and it is also possible that on one intersection there are emergency vehicles coming from all possible sides [26].

3.6 Set the Timer Accordingly

The entire concept used to set the timer is based on the concept of variance, which states that:

Variance is a method of measuring how far a group of numbers extends. This explains how a random variable differs from its estimated value. The difference is

Fig. 18 Computing shortest path using Dijkstra's algorithm

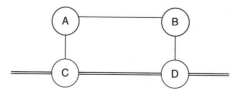

the average of the squares of the difference between the person (observed) and the value devalued [26].

Example
- List = [40,42,48,50]

Variance from the above list will be 17, which is a low variance as compared, thus providing less time for each side's green signal.
- List = [1,2,4,50]

This list will end in a high variance of 427.1875 as compared, and hence performance-wise timing will be allotted.
- List = [10,20,30,40]

This list will result in moderate variance of 125 as compared, and a timer of a maximum of 40 s should be allocated (Fig. 19).

Allocating the Timer on the above Concept
List = [T, B, R, L]

T: represents traffic on the top
B: represents traffic on the bottom
R: represents traffic on the right
L: represents traffic on the left

T, B, R, L are integer values of the cars on the corresponding lane. These values can be altered when a certain IoT chip is notified that an emergency vehicle will be approaching the intersection.
Say:

$x_1 \rightarrow$ number of cars on the top side
$x_2 \rightarrow$ number of cars on the bottom side
$x_3 \rightarrow$ number of cars on the right side
$x_4 \rightarrow$ number of cars on the left side
$m \rightarrow$ number of emergency vehicles on the right side
$n \rightarrow$ number of emergency vehicles on the bottom side
$w \rightarrow$ weight of the emergency vehicle

Initial list $\rightarrow [x_1, x_2, x_3, x_4]$
Modified list $\rightarrow [x_1, x_2+n*w, x_3+m*w, x_4]$

Fig. 19 Pictorial representation
of a junction

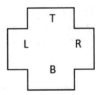

Fig. 20 Pseudocode for variance calculation

```
VarianceVal=Variance(List)
    if VarianceVal<100:
        return 25
    elif VarianceVal<200:
        return 40
    elif VarianceVal<400:
        return 50
    else:              #maximum 650
        return 60
```

The weight, w, is added on the respective sides so that the emergency vehicle can be prioritized. The value of w can be preset to the emergency factor of the vehicle. Emergency factor, w, can be set to 5, 10, or 15 accordingly but should be set to greater than one because one is the count of a single vehicle in a lane [27] (Fig. 20).

4   Conclusion and Future Work

The use of machine learning and image processing to process the headcount of vehicles and track emergency vehicles defines a perfect solution to the problem. By giving the preference to the emergency vehicles, the traffic lights can be altered but it is not the only factor that should affect the traffic light timings. Adjusting the traffic flow, i.e., the solution to the congestion problem, is also the much-needed solution to make things work in a more synthesized manner. Hence, we can conclude that the model is capable of finding a more optimal solution for the required problem by seeing the larger aspect of the problem [28].

The IoT chip used to compute the change in time could be further used to record background data such as the traffic congestion and also could be used to generate the e-challan in case. The model could be further modified to detect the offending behavior of any vehicle and also to get it working with various public departments for the control of road traffic in any condition whatsoever.

References

1. Y.-S. Huang, J.-Y. Shiue, J. Luo, A traffic signal control policy for emergency vehicles preemption using timed petri nets. IFAC-PapersOnLine **48**(3), 2183–2188 (2015)
2. J. Soh, B.T. Chun, M. Wang, Analysis of road image sequences for vehicle counting, in *1995 IEEE International Conference on Systems, Man and Cybernetics. Intelligent Systems for the 21st Century*, vol. 1, (IEEE, 1995), pp. 679–683
3. V.R. Vijaykumar, P.T. Vanathi, P. Kanagasabapathy, Fast and efficient algorithm to remove gaussian noise in digital images. IAENG Int. J. Comput. Sci. **37**(1), 300–302 (2010)

4. A.B. Dorothy, S.B. Kumar, J.J. Sharmila, IoT based home security through digital image processing algorithms, in *2017 World Congress on Computing and Communication Technologies (WCCCT)*, (IEEE, 2017), pp. 20–23
5. Smart traffic light control system: Bilal Ghazal faculty of sciences IV Lebanese University (UL) Zahle, Lebanon bilal.ghazal@ul.edu.lb, Khaled EIKhatib School of Engineering Lebanese International University (LIU) Khyara, Lebanon, Khaled Chahine Beirut, Mohamad Kherfan Kab-Elias, at ISBN: 978-1-4673-6942-8/16/$31.00 2016 IEEE
6. A. Frank, Y.S. Al Aamri, A. Zayegh, IoT based Smart Traffic density Control using Image Processing, in *2019 4th MEC International Conference on Big Data and Smart City (ICBDSC)*, (IEEE, 2019), pp. 1–4
7. A.B. Dorothy, S.B. Kumar, J.J. Sharmila, IoT based Home Security through Digital Image Processing Algorithms, in *2017 World Congress on Computing and Communication Technologies (WCCCT)*, (IEEE, 2017), pp. 20–23
8. Role of IoT technology in agriculture: a systematic literature review Muhammad Shoaib Farooq 1, Shamyla Riaz 1, Adnan Abid 1, Tariq Umer 2 and Yousaf Bin Zikria 3,* 1 Department of Computer Science, University of Management and Technology, Lahore 54770, Pakistan; shoaib.farooq@umt.edu.pk (M.S.F.); S2017108003@umt.edu.pk (S.R.); adnan.abid@umt.edu.pk (A.A.) 2 Department of Computer Science, COMSATS University Islamabad, Lahore 45550, Pakistan; tariqumer@cuilahore.edu.pk 3 Department of Information and Communication Engineering, Yeungnam University, Gyeongsan 38541, Korea * Correspondence: yousafbinzikria@ynu.ac.kr
9. Image Processing of IoT Based Cherry Tomato Growth Monitoring System Nadya Ayu Anugraheni Physics Engineering School of Electrical Engineering Telkom University Bandung, Indonesia nadyaayu15.na@gmail.com Asep Suhendi Physics Engineering School of Electrical Engineering Telkom University Bandung, Indonesia Hertiana Bethanigtyas Physics Engineering School of Electrical Engineering Telkom University Bandung, Indonesia hertiana@telkomuniversity.ac.id – 2019 6th International Conference on Instrumentation, Control, and Automation (ICA) Bandung, Indonesia, 31 July – 2 August 2019.
10. IoT based Smart Traffic density Control using Image Processing Anilloy Frank Dept.of Electronics and Communication Engineering Middle East College Oman anilloy@mec.edu.om Yasser Salim Khamis Al Aamri Dept.of Electronics and Communication Engineering Middle East College Oman pg16F1637@mec.edu.om Amer Zayegh Dept.of Electronics and Communication Engineering Middle East College Oman amer@mec.edu.om
11. Design of IoT Based Intelligent Parking System Using Image Processing Algorithms Sagar Rane Aman Dubey, Tejisman Parida Assistant Professor, Computer Engineering Students, Computer Engineering Army Institute of Technology, Pune Army Institute of Technology, Pune – Proceedings of the IEEE 2017 International Conference on Computing Methodologies and Communication (ICCMC)
12. S. Kumar Janahan, M.R.M. Veeramanickam, S. Arun, K. Narayanan, R. Anandan, S. Javed Parvez, IoT based smart traffic signal monitoring system using vehicles counts. Int. J. Eng. Technol. **7**, 221–309 (2018)
13. P.V. Ashok, S. Siva Sankari, V.M.S. Sankaranarayanan, IoT based traffic signalling system. Int. J. Appl. Eng. Res. **12**(19), 8264–8269 (2017)
14. V. Srinivasan, Y. Priyadharshini Rajesh, S. Yuvaraj, M. Manigandan, Smart traffic control with ambulance detection, in *IOP Conference Series: Materials Science and Engineering*, vol. 402, (2018), p. 012015
15. Anomaly Detection, Analysis and Prediction Techniques in IoT Environment: A Systematic Literature Review MUHAMMAD FAHIM AND ALBERTO SILLITTI Institute of Information Systems, Innopolis University, 420500 Innopolis, Russia Corresponding author: Muhammad Fahim (m.fahim@innopolis.ru) IEEE Access.
16. L. Unzueta, M. Nieto, A. Cortés, J. Barandiaran, O. Otaegui, P. Sánchez, Adaptive multicue background subtraction for robust vehicle counting and classification. IEEE Trans. Intell. Transp. Syst. **13**(2), 527–540 (2011)

17. R. Sundar, S. Hebbar, V. Golla, Implementing intelligent traffic control system for congestion control, ambulance clearance and stolen vehicle detection. IEEE Sens. J. **15**(2), 1109–1113 (2014)
18. Priority Management of Emergency Vehicles at Intersections Using Self-organized Traffic Control Wantanee Viriyasitavat and Ozan K. Tonguz Carnegie Mellon University, Dept. of Electrical and Computer Engineering, Pittsburgh, PA 15213-3890, USA
19. https://opencv-python-tutroals.readthedocs.io/en/latest/py_tutorials/py_video/py_bg_subtraction/py_bg_subtraction.html
20. https://www.learnopencv.com/otsu-thresholding-with-opencv/
21. https://docs.opencv.org/3.1.0/d7/d4d/tutorial_py_thresholding.html
22. https://docs.opencv.org/3.1.0/d9/d61/tutorial_py_morphological_ops.html
23. A. Di Febbraro, D. Giglio, N. Sacco, Urban traffic control structure based on hybrid petri nets. IEEE Trans. Intell. Transp. Syst. **5**(4), 224–237 (2004)
24. S. Rane, A. Dubey, T. Parida, Design of IoT based intelligent parking system using image processing algorithms, in *2017 International Conference on Computing Methodologies and Communication (ICCMC)*, (IEEE, 2017), pp. 1049–1053
25. https://vaibhavsethia.github.io/Smart-Traffic-Light/: Vaibhav Sethia's GitHub model https://github.com/vaibhavsethia/Smart-Traffic-Light
26. D.K. Mahato, S. Yadav, G.J. Saxena, A. Pundir, R. Mukherjee, Image processing and IoT based innovative energy conservation technique, in *2018 4th International Conference on Computational Intelligence & Communication Technology (CICT)*, (IEEE, 2018), pp. 1–5
27. M. Lei, D. Lefloch, P. Gouton, K. Madani, A video-based real-time vehicle counting system using adaptive background method, in *2008 IEEE International Conference on Signal Image Technology and Internet Based Systems*, (IEEE, 2018), pp. 523–528
28. M. Dinesh, K. Sudhaman, Real time intelligent image processing system with high speed secured internet of things/image processor with IOT, in *2016 International Conference on Information Communication and Embedded Systems (ICICES)*, (IEEE, 2016), pp. 1–5

Combining Image Processing and Artificial Intelligence for Dental Image Analysis: Trends, Challenges, and Applications

M. B. H. Moran, M. D. B. Faria, L. F. Bastos, G. A. Giraldi, and A. Conci

1 Introduction

In the last decades, digital image examinations have been introduced in dental practice, and nowadays, they constitute a prevalent tool employed in the diagnosis of oral diseases. Digital sensors have shorter radiation exposure time than analog radiographs. Moreover, digital radiography provides high-quality images. Its uses have been increasing in clinical practice and scientific researches, facilitating the application of computer methods to process and analyze examinations.

The most common examinations in dental practice are the intraoral (periapical, bitewing, and occlusal) and extraoral (especially the panoramic) radiographs and the cone-beam computed tomography. Each one of these types of imaging focuses

M. B. H. Moran (✉)
Instituto de Computação, Universidade Federal Fluminense, Niterói, Brazil

Departamento de Diagnóstico e Cirurgia, Faculdade de Odontologia, Universidade do Estado do Rio de Janeiro, Rio de Janeiro, Brazil
e-mail: mhernandez@id.uff.br

M. D. B. Faria
Departamento de Diagnóstico e Cirurgia, Faculdade de Odontologia, Universidade do Estado do Rio de Janeiro, Rio de Janeiro, Brazil

Faculdade de Odontologia, Universidade Federal do Rio de Janeiro, Rio de Janeiro, Brazil

L. F. Bastos
Departamento de Diagnóstico e Cirurgia, Faculdade de Odontologia, Universidade do Estado do Rio de Janeiro, Rio de Janeiro, Brazil

G. A. Giraldi
Laboratório Nacional de Computação Científica, Petrópolis, Brazil

A. Conci
Instituto de Computação, Universidade Federal Fluminense, Niterói, Brazil

© Springer Nature Switzerland AG 2022
P. Johri et al. (eds.), *Trends and Advancements of Image Processing and its Applications*, EAI/Springer Innovations in Communication and Computing,
https://doi.org/10.1007/978-3-030-75945-2_4

on different anatomical structures and is used for different purposes. Their acquisition processes also differ one from another. One of the topics discussed in this chapter is the principles of the image formation and acquisition processes behind those techniques.

According to Abdalla-Aslan [1], computer methods, especially those that include artificial intelligence (AI), can be used to improve the accuracy and consistency of diagnosis. Various AI solutions for oral radiology have emerged in the last few years. Several previous works present efforts to automate the identification and evaluation of oral diseases in image-based exams, which would reduce possible errors related to experts' subjectivity [1].

Although most works use traditional image processing methods, recently, machine learning algorithms and even convolutional neural networks (CNNs) have shown promising results for this problem. According to Schwendicke et al. [2], dental imaging presents an excellent potential for image processing solutions since diagnostic imaging is an essential part of dentistry. In other words, image evaluation already consists of an important step in the diagnosis of several oral diseases.

In European countries, most of the radiographs acquired consist of oral image examinations [2]. It is estimated that around 250–300 dental images are acquired per 1000 individuals [2]. AI-based solutions tend to be suitable for a wide range of image applications, including oral ones. This chapter also briefly presents their main principles.

The use of AI-based image techniques tends to increase the effectiveness of diagnosis and lower costs by eliminating routine tasks [3]. Consequently, in the last years, the number of works proposing solutions on the use of such techniques has been increasing.

The main areas of dentistry for which AI-based image processing techniques were applied are cariology, endodontics, periodontology, orthodontics, and forensic dentistry [2].

The most popular applications are segmentation, detection, classification, and their combinations [2]. The detection is related to carious lesions. The anatomical structures for classification include teeth, jaw bone, skeletal landmarks, and biofilm classification. Classification considers the endodontic treatment conditions and even their results. The detection and classification evaluate periodontal inflammation, bone loss, and facial features [2].

The tasks involved in those applications are localization and measurement of anatomic structures, diagnosis of osteoporosis, classification and segmentation of maxillofacial cysts or tumors, identification of alveolar bone resorption, classification of periapical lesions, diagnosis of multiple dental diseases, and classification of tooth types [4]. Other applications less explored are identification of root canals, diagnosis of the maxillary sinusitis, identification of inflamed gum, identification of dental plaque, detection of dental caries, and classification of the stages of the lower third molar [4]. This chapter also discusses some of the main application problems for AI-based image processing. To demonstrate the feasibility of using the presented techniques, three of the mentioned applications (identification of periodontal diseases, detection of dental caries, and radiograph image enhancement) are selected for practical exemplification.

Although there is a great potential for the use of AI-based image processing techniques in dental imaging, there are several challenges to be overcome in this context. Among these are the lack of available data, the subjectivity of oral diseases' diagnosis, the lack of diagnostic standards, the complexity of some oral diseases, and the resistance from dentists to include computational tools in their routine. All these aspects are discussed in this chapter, as well.

2 An Overview on Digital Dental Imaging

This section discusses the principles of image formation and representation for oral radiographs and presents the most common imaging exams used in dentistry.

2.1 X-Ray Images

Medical radiographs consist of a type of biomedical data from particles' interactions in the electromagnetic x-ray spectrum, which consider very short wavelengths. The range that covers x-ray photon energies is delimited from 10 keV (1.6×10–15 J) to 100 keV, *i.e.*, wavelengths ranging from 0.124 to 0.0124 nm [5]. Basically, the devices used to obtain radiographic exams are constituted by an x-ray source and a detector, also called receptor (Fig. 1), which can be a film or a digital device.

The visual interpretation of radiographic exams is based on the radiodensity concept. When x-ray photons are irradiated from an x-ray source to an object (or biological tissue) composed of an impenetrable material, the amount of radiation

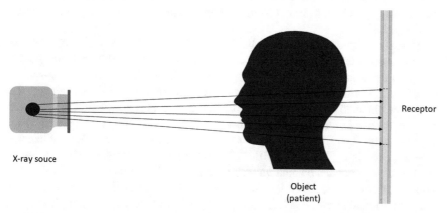

Fig. 1 Main configuration of the patient and devices in the acquisition of radiography

reaching the detector device (or film) is low since the object absorbs most of the x-ray photons. Consequently, its respective projection in the final based radiography is light. This property of materials of retaining x-rays and producing light is called radiopacity. On the other hand, when x-ray photons are irradiated in a material that promotes its propagation, the amount of radiation that reaches the detector device is high, resulting in a respective dark projection in the radiographic image. This property is called radiolucency [5, 6].

Attenuation is the main physical property used in image formation for most conventional x-ray machines and computed tomography scan systems. It is defined as the difference in the amount of x-ray energy emitted by the transmitter and the energy received by the receptor (digital sensor or film) after transposing an object (patient's biological tissue) during the examination.

For dental radiographs, such differences related to the attenuation values can result in four aspects: (1) coherent scattering (when incident photons scatter from outer electrons), (2) photoelectric absorption (when incident photons eject their inner electrons and fade away, releasing photons of a characteristic type), (3) Compton scattering when incident photons eject outer electrons, and (4) other possible scatterings. Attenuation values are commonly expressed in Hounsfield units (HU) that are based in the water's attenuation, so the resultant HU value for a specific tissue can be defined as:

$$HU_{tissue} = 1000 \times \frac{\mu_{tissue} - \mu_{water}}{\mu_{water}} \tag{1}$$

where μ_{water} and μ_{tissue} are the water's and tissue's attenuations.

After the exam execution, its results can be represented as images. The analog approach for that is achieved by using film imaging and its processing as in analogical photography. Image formation based on film imaging uses a sensitive layer that is modified by x-ray photons, through oxidation, proportionally to the amount of radiation exposition [5]. The film is then chemically processed to produce a grayscale image, reflecting the x-ray opacity of each tissue in a continuous range (i.e., this analog approach is similar to the principles applied for the former analogical film imaging).

Digital radiographic application has been increased in the last years, allowing the use of several computer-based processing, like those presented here. Radiography acquired using the digital approaches presents differences from the analog-signal-based ones, starting from the way they are represented. In digital radiography, the measures obtained by the acquisition process are spatially distributed in a discrete way, represented in a digital file and used in a matrix-like structure, defined by its resolution that is the number of rows and columns of such a matrix [5, 6]. When the electronic receptor, used in digital devices, absorbs the x-ray photons that go through the object, it generated a small voltage for each pixel, which is proportional to the volume of photons received by each device position. After that, a process called analog-to-digital conversion (ADC) is performed. It consists of defining ranges for the voltage's values obtained in a way that the pixels whose values are in a defined range are grouped together, and then the same digital value is assigned to a point [5, 6], forming the digital image. A visualization tool (as a computer monitor) reads these values and assigns a corresponding gray shade for each one to display the matrix as a grayscale image.

According to their physical properties, electronic receptors can be divided into solid-state detectors or photo stimulus phosphor detectors. The first uses solid semi-conducting materials to gather the charge generated by x-ray photons. The second uses photostimulable phosphor plates to absorb, store x-ray energy, and further release it as light after being stimulated by another light that presents an appropriate wavelength [6].

In order to obtain accurate measures and, consequently, higher-quality images, radiation doses should be adjusted. Nevertheless, a larger exposition to radiation poses a risk for the patient's health. For oral radiography, standard radiation ranges were proposed in order to obtain the highest quality preserving the patient's safety.

Usual radiography is an exam that consists of a planar projection of a 3D scene. This scene is composed of the patient's biological tissues and anatomical structures. The way the patient and the device are positioned changes with the exam focus and, consequently, the resultant projected image. Many configurations are defined to cover the different anatomic parts. The next section discusses the most common oral ones, including the positioning of the elements to achieve the images focusing on different structures according to the visualization objective.

2.2 Intraoral Radiographs

As suggested by the term, the acquisition process of intraoral radiographs involves positioning part of the device inside the patient's mouth. There are three main types of intraoral radiographs used in dental imaging: periapical views, bitewing views, and occlusal views. Figure 2 shows them.

To cover all the dental arches of a healthy adult patient, a set of 17 periapical views and 4 bitewing views is required for most cases [6]. Periapical views (Fig. 2(a)) cover the teeth's crowns, roots, and surrounding bones. Figure 3 shows

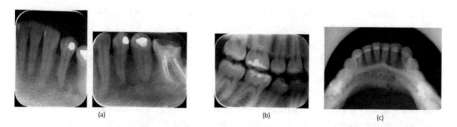

Fig. 2 Examples of (**a**) periapical projections, (**b**) bitewing projections, and (**c**) occlusal projections. Occlusal view by Coronation Dental Specialty Group under CC BY 3.0 via Wikimedia Commons

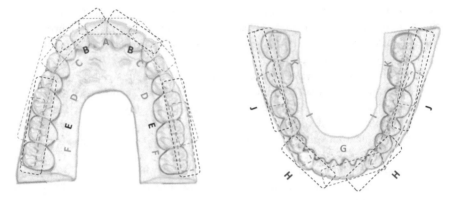

Fig. 3 Target teeth of periapical projections A to K

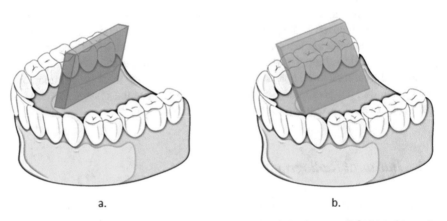

a.   b.

Fig. 4 Receptor positioning in the periapical radiography: (**a**) paralleling and (**b**) bisecting angle

the distribution of a complete set of periapical projections and the teeth they respectively cover. Projection A refers to the maxillary central incisors, projection B refers to the maxillary lateral incisors, projection C refers to the maxillary canines, projection D refers to the maxillary premolars, projection E refers to the maxillary molars, projection F refers to the maxillary distomolar, projection G refers to the mandibular centrolateral incisors, projection H refers to the mandibular canines, projection I refers to the mandibular premolars, projection J refers to the mandibular molars, and projection K refers to the mandibular distomolars.

Two projection techniques are mostly used in periapical radiograph acquisitions: paralleling and bisecting angle (Fig. 4).

The parallel periapical radiographs tend to result in an image with less distortion and are commonly recommended for digital imaging. This technique consists of positioning the x-ray receptor as parallel as possible to the dental arches inside the patient's mouth, so the projection is obtained orthogonal to the teeth and the receptor plane [6]. Considering a plane that approximates the surface of few consecutive teeth (as the maxillary central incisors), the main idea of the parallel technique is positioning the receptor as a parallel plane of the mentioned teeth plan, so the x-ray hit them directly, in a perpendicular direction (Fig. 4(a)). This better reflects the teeth's true anatomical characteristics, reducing distortion in the acquisition.

The bisecting-angle technique is only used when the parallel technique cannot be applied due to large rigid sensors or the patient's anatomy. It is based on the geometric principle that states that two triangles are equal if they have two equal angles and share completely one of their sides. In this projection, the receptor is positioned as close as possible to the internal part of the dental arch, i.e., the lingual surface of the teeth. If the exam focuses on the mandibular teeth, the receptor must be held from the bottom by the palate. If the exam focuses on the maxillary teeth, the receptor must be held on top by the floor of the mouth (Fig. 4(b)). Holding instruments (or the patient's fingers) are used to promote the receptor's perfect positioning for both techniques and the x-ray emission direction is adjusted in a proper manner.

Bitewing views (Fig. 2(b)) are also called interproximal views. They cover the coronal portions of the maxillary, mandibular molars, and premolars in a single image. Four of these projections are acquired with the periapical views covering all arches: two for the premolars (Fig. 5a) and two for the molar teeth (Fig. 5b).

Occlusal projection is another possible intraoral radiograph. It covers a wide part of the dental arches. This exam is mostly used when the patient's mouth cannot hold the periapical receptors. As suggested by the term, the receptor is placed in the occlusion plane. In occlusal acquisitions, the receptor is located between the occlusal surfaces of the teeth. The most common occlusal views are anterior maxillary occlusal projection (Fig. 6a), cross-sectional maxillary occlusal projection (Fig. 6b), lateral maxillary occlusal projection (Fig. 6c), anterior mandibular occlusal projection (Fig. 6d), cross-sectional mandibular occlusal projection (Fig. 6e), and lateral mandibular occlusal projection (Fig. 6f).

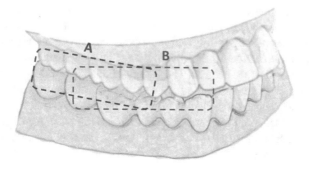

Fig. 5 Target teeth of the bitewing projections

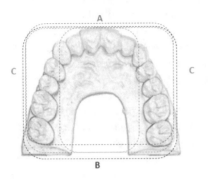

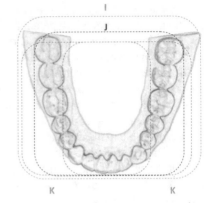

Fig. 6 Target teeth of the occlusal projections

Fig. 7 Example of
panoramic radiography.
Paronamic radiograph by
Umanoide under CC via
Unsplash

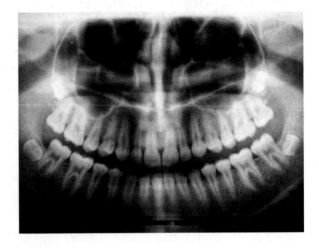

2.3 Extraoral Radiographs

Extraoral radiographs, as suggested by the term, consist of radiographic exams that
do not include introducing part of the device into the patient's mouth. The most
widely used extraoral image in dental practice is the panoramic view presented in
Fig. 7. The panoramic view covers all of the maxillary and mandibular dental arches
and a wide part of the face (Fig. 7). The quality of this kind of image can be considered lower than the intraoral radiographs since it promotes geometric distortions,
moving shadows, and even the inclusion of overlapping effects due to the presence
of other anatomical structures in the dental arch proximity, as the neck bones. It is
mainly recommended for initial evaluations and for cases in which intraoral radiographs cannot be acquired [6].

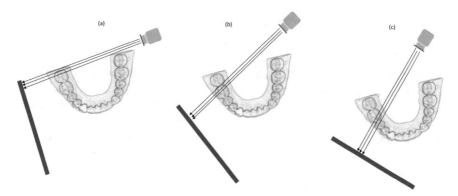

Fig. 8 Representation of the acquisition process in a panoramic view

For panoramic acquisition, the object of interest (mouth of the patient) is positioned in the plane (image layer) in a central point in relation to the x-ray source and receptor, which are on opposite sides. Then receptor and x-ray source move simultaneously. The panoramic image is formed dynamically, that is, its acquisition is made during the device movement, so each part of the image corresponds to a different position and time. To analyze this, due to such a dynamic capture, the receptor movement has to be considered, as well as the x-ray source's position and also the part of the mouth which is currently on focus.

Figure 8 shows the position of the device at three different times of the panoramic acquisition. Note that Fig. 8(a) corresponds to the first tooth of the acquisition process, so the part of the receptor that is directly receiving the x-rays emitted by the source corresponds to this part of the mouth. As the device continues, the acquisition process covers the rest of the dents and the receptor also moves to receive the x-ray of the corresponding part of the mouth (Fig. 8(b) and (c)). Note that the receptor moves close to the patient's teeth arches while the x-ray source moves behind the patient's neck. The receptor is intentionally positioned in this fashion because the structures close to the receptor are better projected in the resulting image. Due to the projection principles of these images, the structures that are close to the x-ray source are projected in the formed image in a way that it appears with magnification, resulting in deformed results and blur [6].

Extraoral radiographs also include several other projections. The most used ones are lateral skull projection (lateral cephalometric projection), submentovertex (base) projection, Waters' projection, posteroanterior skull projection (posteroanterior cephalometric projection), reverse Towne projection (open mouth), and mandibular oblique lateral projections. These projections are mostly used in orthodontics and cephalometric landmark identification [6].

2.4   Computed Tomography and Cone-Beam Computed Tomography

Traditional computed tomography (CT), also called fan-beam tomography, acquires three-dimensional images by irradiating x-ray beams linearly. In order to cover a 3D object, the source and the receptor must surround a central point of the object plane, as shown in Fig. 9. When a rotation of 360° is completed, both source and receptor translate in the direction of the plane's normal axis, covering all the volume (object) [7]. Each rotation of the source results in a planar slice. As the source moves along the axial direction, new slices are obtained. After the end of the acquisition process, the slices are then computationally processed, creating a 3D digital volume. For a pair of consecutive slices, the values are interpolated to fill the region between them, resulting in a continuous-like representation. The number of slices depends mainly on the device's characteristics. Over time, CT acquisition process has improved, especially concerning the number of slices and the way that the x-ray source's rotation and axial displacement are performed. Nevertheless, the current CT devices still follow these principles.

In dental imaging, the most prevalent 3D image is the cone-beam computed tomography (CBCT). In the acquisition process of CBCT, 3D images are acquired in only one rotation of the x-ray source. The beam used in this process has a cone format, so in each step of the rotation, a complete 2D projection is acquired at once (Fig. 10). There is no need for axial displacements of the x-ray source [7]. So the number of slices depends mainly on the digital receptor discretization. For CBCT, the acquisition produces an entire discrete 3D volume, with voxels' sizes according to the receptor pixel size [7].

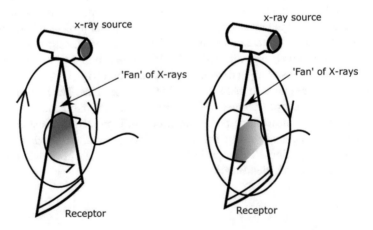

Fig. 9 Representation of the acquisition process of CT at two different positions

Fig. 10 Representation of
the acquisition process
of CBCT

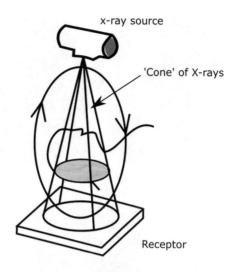

x-ray source

'Cone' of X-rays

Receptor

3 Image Processing and Artificial Intelligence for Dental Image Analysis

This section describes the most common image processing tasks involving artificial intelligence techniques and their real application for dental imaging. In addition to image enhancement (which can be considered as an interesting application for dental imaging), it also discusses the most prevalent applications of the presented techniques for two dentistry sub-areas: periodontology and cariology. Some previous works in literature that present solutions for problems in the area are considered, as well. Moreover, practical examples applying AI and image processing for tasks involving classification, detection, and image enhancement are analyzed.

3.1 Artificial Intelligence Techniques for Image Processing

Digital image processing (IP) is a field of computer science and signal processing that studies digital signals presenting two-dimensional (2D) structures. Radiography is a type of biomedical imaging where the radiographic devices themselves apply some of these techniques after the images' acquisition and before their storage [6]. A wide range of available IP processing techniques can be employed to improve, analyze, and extract information from oral radiographs. Researchers can apply IP techniques as it better suits their objectives. Four techniques are mostly related to dental imaging applications: enhancement, segmentation, identification, and classification. Figure 11 illustrates them.

The enhancement task consists of processing the image to improve its quality concerning noise, resolution, edge definition, etc. The objective of the segmentation

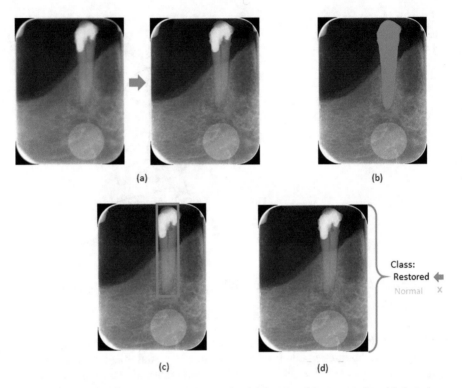

Fig. 11 Example of image processing tasks: (**a**) enhancement, (**b**) segmentation, (**c**) detection, and (**d**) classification

task is to identify an object in the image, determining its exact boundaries and isolating it from the rest of the image. Figure 11(b) shows an example of this task (tooth segmentation). The identification task focuses on determining the region of the image that encloses an object, for example, tooth detection (Fig. 11(c)). Finally, the classification task consists of analyzing the entire image and its visual patterns to associate it to a specific class, as in the example in Fig. 11(d), in which a tooth is classified as normal or restored.

In the last years, the use of artificial intelligence (AI) techniques as a support to traditional IP has been increasing, leading to results that demonstrate their potential. Convolutional neural networks (CNNs) are an essential part of AI being the basis of most AI-based algorithms for IP nowadays. CNNs consist of a specialized kind of intelligent algorithm for processing data that present a grid-like topology as digital images.

The IP convolution operation is the base of CNN algorithms. Convolution consists of transforming an input image using a kernel to achieve a feature map as output. More specifically, given two matrices, with the same numbers of elements, i.e., $n \times$, one named kernel and the other being a part of the image to be convoluted, then convolution consists of multiplying correspondent position and adding them to

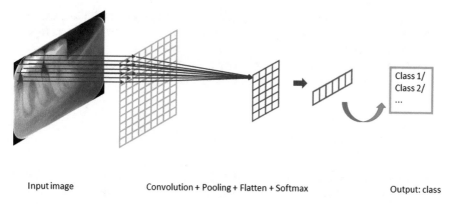

Input image　　　　　Convolution + Pooling + Flatten + Softmax　　　　　Output: class

Fig. 12 Schematic representation of an input, a simple CNN, and an output

obtain a value that is used to compose the output for each position that the kernel can cover on the image. Figure 12 exemplifies the convolution and other operations that can be combined to compose a simple CNN to perform a classification task.

The output named feature map can be processed by another convolution or even other operations. Among the most used other types of operation are size reduction (or pooling), dimension reduction resulting in a 1D vector (or flatten), and mapping of the position to a class previously defined (or softmax).

Note that different CNN architectures can be achieved by combining convolutions with those operations in various manners, modifying the number of layers and even the way the operations are organized. Some architectures proved to be efficient for a wide range of applications receiving particular names as the ResNet and Inception. The most straightforward application of CNNs is for classification tasks.

Some detection and segmentation tasks can be modeled as extensions of the classification. For example, consider a 95 × 20 area of the input that includes a tooth to be detected as in Fig. 11(c). Note that if one divides this image into two sub-images of size 5 × 5 and classifies each sub-image to identify if they present a tooth or not, then the sub-image that encloses the tooth can be identified. Therefore, the region that is a union of each tooth sub-image can be considered as the tooth region resulting in a detection. However, in real cases, the object (tooth in this example) may not be so well positioned, so several different subdivisions must be tested to achieve the segmentation that covers only one complete object. This is the main idea behind most AI-based detection algorithms.

The segmentation task can also be considered as an extension of the classification task, but on a pixel scale, in the sense that each pixel is classified as belonging to the object's area or not (Fig. 11(b)). The main idea behind most AI-based segmentation algorithms is to analyze each pixel, considering their neighborhood, which is defined by a window size, to evaluate if it corresponds to the patterns that characterize the object. In other words, each pixel is classified as being part of the object or not considering a tiny sub-image as input, which is defined by a window that covers its neighborhood.

Actually, the real algorithms that perform these tasks are much more complex than this simple description presented here since they include several layers (operations), but this gives us the main idea of how they work.

Note that all these concepts can be extended to N-dimensional signals, including 3D data, so they can also be applied in tomographs, for example.

3.2 Identify Periodontitis

Periodontal disease (PD) is a consequence of interactions between bacterial biofilm and the host's immune response [6, 8], and differences in the degree of severity and impairment of this disease can be influenced by extrinsic factors, such as smoking, and intrinsic factors, such as diabetes mellitus [9]. PD can be divided into gingivitis and periodontitis [6, 10]. One of the consequences of tissue destruction due to periodontitis is bone loss. Radiographically and clinically, this loss can be observed as an increase in the distance between the enamel-cement junction to the alveolar crest.

The fact that this tissue destruction can be identified radiographically promotes the use of AI-based image processing techniques for this purpose. Recently, Lin et al. [11, 12] proposed the use of deep learning models for alveolar bone loss identification [11] and measurement [12]. The model proposed by Lee et al. [13] focuses specifically on the identification and severity assessment of premolars and molars periodontally compromised. Similarly, the studies of Carmody et al. [14] and Mol et al. [15] aim to classify periapical lesions according to their extent. A considerable part of the works that focus on identifying periodontitis/periapical diseases uses panoramic radiographs. For example, Ekert et al. [16] used convolutional neural networks (CNNs) to detect apical lesions on panoramic dental radiographs. The implemented network, a custom-made seven-layer deep neural network, achieved a sensitivity value of 0.65, a specificity value of 0.87, a positive predictive value of 0.49, and a negative predictive value of 0.93. Krois et al. [17] applied a seven-layer deep neural network to detect PBL on panoramic dental radiographs. The classification accuracy of the CNN was 0.81, and the sensitivity and specificity were 0.81 and 0.81, respectively.

Classifying Approximal Bone Loss in Periapical Radiographs Identification of periodontal diseases is a common application area for AI-based image processing, as exposed previously in this section. Intraoral radiography, especially periapical exams, is an important tool for identifying these anomalies, facilitating their diagnosis, treatment, and prognosis [18]. Next, this section demonstrates the use of some AI and image processing techniques to pre-process and classify interproximal regions in periapical examinations according to the presence of proximal bone loss. For that, a brief evaluation of the use of two CNNs architectures is performed, specifically ResNet and Inception networks, to demonstrate how different architectures can influence the quality of the final results.

This experiment used 1079 interproximal regions manually extracted from 467 different periapical radiographs. All images are in grayscale, in "jpeg" format. This experiment is focused on a classification task. Therefore, the region extraction was performed manually. The next section covers a detection task to automatically extract the regions of interest from oral radiographs using image processing techniques.

Firstly, an adaptive histogram equalization [19] was applied to the periapical images in order to increase their quality. The adaptive histogram equalization is an image processing technique used to improve contrast in images and enhance their details. It adjusts the image contrast by considering its most frequent tonalities. The process is similar to the original histogram equalization; however, it considers parts of the images rather than the entire image, allowing it to create different histograms and use them to calculate the equalization [19]. The main idea of this technique is to define a neighborhood window to be considered in the histogram of the transformation function for each pixel. In this experiment, after some initial testing, an 8×8 window was selected. As in the ordinary histogram equalization, the transformation function of the adaptive histogram equalization is proportional to the cumulative distribution function (CDF) of the pixel values in the neighborhood [19].

Experts marked the regions of interest (ROIs) in each exam. The ROIs cover the areas that can be affected by bone loss. These regions consist of interproximal (between two teeth) areas, limited superiorly by the enamel-cement junctions and inferiorly by the alveolar crests. To be used as input of a convolutional neural network for the proposed classification task, all these data must be labeled, i.e., for each case/image an associate class must be assigned to it by experts. This process is called data labeling and can be performed using several auxiliary tools. This example used the labeling tool named DataTurks (available at https://dataturks.com/). Two experts annotated the exams' ROIs, using bounding boxes, denoting which of them present any bone loss and which do not. They are experienced dentists specialized in oral radiology. They annotated 1079 regions: 388 with no lesions and 691 with bone loss (no differences between experts' annotations).

In order to prepare the data for the classification task, this data must be organized into three different sets: training, validation, and test sets. This process is called dataset split. The test dataset was formed by 52 samples of each class randomly selected. The remaining images underwent a data augmentation process based on horizontal and vertical flips. After that, the 639 remaining annotated regions with vertical bone loss provided 1278 images, using only horizontal flips. The remaining 336 images of healthy regions provided 1344 images, using both horizontal and vertical flips. In that way, the CNNs' training and validation sets are formed by these 2622 images.

Finally, the actual classification task is performed. As mentioned, this example includes an evaluation of two different CNNs for the classification task in order to compare which is the most appropriate to the proposed problem. This experiment included two architectures that demonstrate good performance for a wide range of applications: ResNet and Inception architectures. The ResNet architecture used in this work has 50 layers in total. It is composed of several stacked blocks, called residual units. Such units consist of two convolutional layers and two activation functions [20]. On the other hand, the Inception architecture is formed by blocks called Inception modules [21], consisting of a combination of convolutional layers with different kernel sizes and a pooling layer. This study used the official Keras ResNet and Inception implementations. The data processing performed in this work used the Python language and the scikit-image library. The parameters used for training the CNNs are outlined in Table 1.

The CNNs' training used the backpropagation algorithm and included 180 epochs. For each epoch, we checked the accuracy and loss values. Each epoch corresponds to one time in which CNN weights are updated considering all elements of the training dataset. The models used in this example were previously pre-trained using the ImageNet dataset [25] to obtain better initial weight values.

An important measure to be considered in the evaluation of CNNs in a classification task is test accuracy (proportion of cases properly classified by the considered model). Other measures are sensitivity (recall), specificity, precision (positive predictive value, PPV), and negative predictive value (NPV) [26]. In this example, such measures are based on:

- True negatives (TN) – regions correctly classified as healthy
- True positives (TP) – regions correctly classified as regions with bone loss
- False negatives (FN) – regions with bone loss incorrectly classified as healthy
- False positives (FP) – healthy regions incorrectly classified as regions with bone loss

In that way, the mentioned measures are defined as sensitivity $= \dfrac{TP}{FN+TP}$, specificity $= \dfrac{TN}{FP+TN}$, precision $= \dfrac{TP}{FP+TP}$, and negative predictive value $= \dfrac{TN}{FN+TN}$ [26].

Most evaluations also include the receiver operating characteristic (ROC) and the precision-recall (PR) curves [26]. In this example, all measures were calculated using Python and the scikit-learn library.

At the end of the training process, the Inception model presented an in-sample accuracy of 0.984 and a validation accuracy of 0.933. On the other hand, the ResNet model had an in-sample accuracy of 0.919 and a validation accuracy of 0.818. Concerning the evaluation based on the test set, the results are shown in the respective confusion matrices (Table 2). Note that the test accuracy (proportion of

Table 1 Hyperparameters used in CNNs' training

Parameter	ResNet	Inception
Optimizer	Momentum [22]	RMSprop [23]
Batchsize	32	32
Learning rate	0.01	0.01
Loss function	RMSE [24]	RMSE [24]

Table 2 Confusion matrices for the ResNet and Inception models

ResNet	n = 104		Predicted	
			Healthy	Vertical bone loss
	Actual	Healthy	38	14
		Vertical bone loss	13	39
Inception	n = 104		Predicted	
			Healthy	Vertical bone loss
	Actual	Healthy	37	15
		Vertical bone loss	4	48

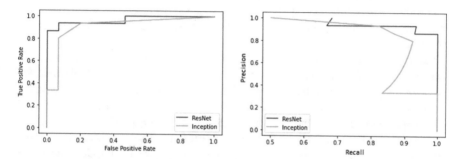

Fig. 13 ROC (left) and PR (right) curves for each model

Table 3 Test results

Measure	ResNet	Inception
PPV (precision)	0.736	0.762
Sensitivity (recall)	0.750	0.923
Specificity	0.731	0.711
NPV	0.745	0.902
AUC-ROC	0.864	0.860
AUC-PR curve	0.868	0.847

examples correctly classified) of the ResNet model was 0.740 and the Inception's was 0.817. Table 3 summarizes the other test measures, and Fig. 13 shows the ROC and PR curves for each model.

Note that the Inception model had the best overall performance (Tables 2 and 3). The lower performance of the ResNet model consists of misclassifications almost equally distributed between both healthy and bone loss classes. On the other hand, the misclassifications for the Inception model are mainly healthy regions incorrectly classified as regions that present vertical bone loss. Finally, the good results of the considered CNNs are denoted by the ROC and PR curves.

3.3   Detection of Dental Caries

Dental caries is a multifactorial oral disease affected by sucrose consumption. It presents a high prevalence [27], and its prevention demands early detection and treatment. Its development depends on the presence of bacteria, especially mutant streptococci, that ferment carbohydrates, resulting in the demineralization of hard dental tissues [28–30]. The accumulation of such bacteria forms what is known as plaque (biofilm) [27]. It initially affects the tooth surface, and after severe demineralization or cavity formation, it can penetrate the hard tissues. Clinically, when it is visible, dental caries presents as a matte white spot (indicating ongoing activity) or an opaque or dark brownish spot (indicating past activity) [6, 31]. Demineralization may extend into the dentin, the enamel, or even the pulp and can destroy the entire tooth structure [30].

Most caries lesions are visible in periapical images. Approximal caries affects the interproximal area between two consecutive teeth. They are generally detected through image examinations, especially bitewing radiographic images, because the positions of such lesions prevent a clinical evaluation. In bitewing images, dental caries appears as a darker area due to their low x-ray absorption [6].

Several previous works have focused on identifying dental caries by examining images such as optical coherence tomography (OCT), periapical radiography, and bitewing images. Although initially, traditional image processing methods were applied in most works [32–35], machine learning algorithms have recently become a more common approach to visual problems, including dental images. Deep convolutional neural network (CNN) algorithms have been used for human oral tissue classification to provide early detection of dental caries [36]. A CNN model analyzes optical coherence tomography (OCT) images of different densities of oral tissues and determines variations related to the demineralization process. That suggests that variations in caries lesion may be identified in other image examinations as well, as previously mentioned.

Deep CNNs have also been applied to the detection and diagnosis of dental caries on periapical radiography images [37]. A pre-trained GoogLeNet Inception v3 model was used to process 3000 periapical radiographs. Three different models were created: a premolar version, a molar version, and a final version for both premolar and molars. These models achieved impressive accuracy results (89.0%, 88.0%, and 82.0%, respectively). Thus, considering the good performance of the presented method, the study showed the feasibility of using a deep CNN architecture to detect and diagnose dental caries.

Bitewing images have also previously been evaluated to identify dental caries stages and potential false diagnoses [38]. In that study, several texture features were extracted from the evaluated images via a gray-level co-occurrence matrix (GLCM). These feature values were processed by an algorithm that combines a logit-based artificial bee colony optimization algorithm with a backpropagation neural network to increase the classification accuracy. The proposed approach achieved an accuracy of 99.16%.

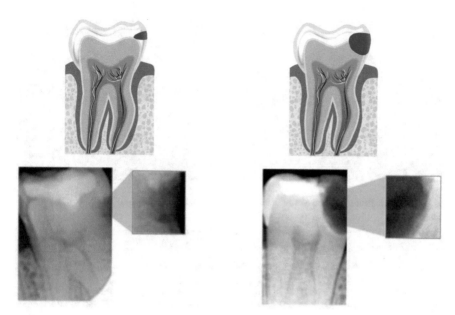

Fig. 14 Tooth stages considering the caries severity: (**a**) representation of tooth with an incipient lesion, (**b**) bitewing image with incipient lesion highlighted, (**c**) representation with an advanced lesion, and (**d**) real example of a bitewing exam with advanced caries

Approximal Dental Caries Detection and Classification in Bitewing Images As previously mentioned in this section, caries detection is a common application area for AI-based image processing. Next, this section demonstrates the use of some AI and image techniques to detect approximal caries in bitewing images and classify them according to their severity. Consider three different caries stages based on their lesion severity: normal (no lesion), incipient (superficial lesion affecting the enamel; Fig. 14a and b), and advanced (lesion affecting a considerable part of the tooth, expanding into the dentin and the pulp; Fig. 14c and d).

The first step to prepare the data for the CNN classification is to detect the teeth in the bitewing radiographs using image processing techniques. Each of the detected teeth was separated, creating individual tooth images. As previously mentioned, tooth detection is a task for which previous works applied deep neural networks, as YOLO and Fast RCNN. Nevertheless, in an ideal scenario, classic image processing techniques may also present a good performance, as demonstrated in this experiment. This experiment excludes cases of dental implants, crowding, and malocclusion. For these cases, deep learning solutions may present better results.

The teeth detection method based on classic image processing techniques has as the first step an equalization operation (Fig. 15) to enhance the details and differentiate between background and tooth areas more easily. This example uses the adaptive histogram equalization. As a result of the equalization process, teeth and background can be more easily differentiated in the images because their tonalities

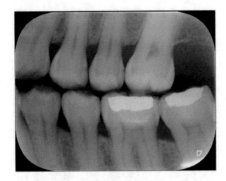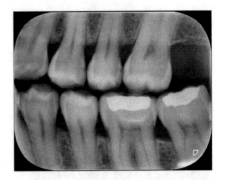

Fig. 15 Application of adaptive equalization: (**a**) original image and (**b**) equalized image

Fig. 16 Pre-processing using morphologic operations: (**a**) thresholded image, (**b**) eroded image, (**c**) open image, and (**d**) dilated image

differ more substantially. Thus, a threshold can be used to transform the original grayscale images into binary images where the background is black and the tooth area is white. This example used the Otsu threshold [39].

Observe that in the resulting binary images, sometimes the gum area is considered as background, and sometimes it is included in the white tooth area due to tonal similarities between the tooth and gum regions. These gum regions are removed using morphological operators [40]. The white areas related to teeth consist of large regions with few holes, while the white areas pertaining to gum are mostly small and irregular and can easily be removed using erosion and opening morphological operations applied consecutively [40]. Considering the thresholded image (Fig. 16a), the next step is to apply erosion using a 130 × 20 rectangle as a structuring element (Fig. 16b). This specific element was chosen after evaluating the gum areas' shapes. The use of smaller elements did not result in the correct elimination of the gum areas. Similarly, the use of larger elements resulted in considerable losses in the identified tooth regions. Furthermore, using a uniform, symmetrical square or circle element did not allow the separation of teeth that are close together.

Next, an opening operation was applied, using a circle with a radius of 20 pixels as the structural element. This operation results in the elimination of the remaining undesirable parts (Fig. 16c). Finally, dilation is applied using a circle with a radius of 15 pixels as a structuring element, which results in the inclusion of the tooth borders in the tooth areas (Fig. 16d).

After removing the gum areas, the binary images are composed of large white areas on a black background. Each area refers to a different tooth. New images of each tooth are created based on the bounding boxes around these areas. Thus, the original image is repeatedly cropped, using the bounding boxes' limits to obtain individual images for each tooth.

A total of 480 different tooth images were extracted from the 112 bitewing radiographs by the described detection method. To be used as an input of a convolutional neural network for the proposed classification task, all data must be labeled, i.e., the lesion severity class must be assigned for each tooth. To obtain the labels for each of the 480 teeth, 2 experts used a labeling tool named DataTurks (available at https:// dataturks.com/) to associate each detected tooth to 1 of the considered classes: healthy, incipient, or advanced. These experts are experienced dentists, and one is specialized in oral radiology. This labeling process pointed out that the set of 480 detected teeth included 305 normal teeth, 113 teeth that present incipient lesions, and 62 teeth that present advanced lesions. There was no discrepancy between their annotations, i.e., they pointed to the same classes for all cases.

The next step in the data preparation for the classification task is the dataset split. The data must be split into training and test sets used to train and evaluate the CNN model, respectively. Fifteen cases of each class are used as a test set, resulting in 45 teeth. The remaining 435 tooth images (divided into 290, 98, and 47 images for normal, incipient, and advanced classes, respectively) underwent a data augmentation process. The data augmentation process consists of creating variation in input images to increase the data volume, which was proved to be essential to achieve good results in deep learning models [41]. This example's data augmentation processes consist of applying rotate and flip operations to the tooth images, creating 1160, 1176, and 1128 sample images for healthy (normal), incipient, and advanced classes, respectively.

Due to the outstanding performances presented by Inception v3 models in prior medical image classification studies, this CNN architecture was chosen to be used in this experiment [42]. The parameters used for training the CNNs are outlined in Table 4.

The models used in this example were previously pre-trained using the ImageNet dataset [25] to achieve better initial weight values. The fine-tuning training process included 11,500 steps and 3 different values (0.1, 0.01, and 0.001) as the initial learning rate to evaluate which of these parameter values would be the most appropriate.

The final accuracy and loss values, considering the training and validation sets, achieved after completing the training process, pointed out that the best Inception model was the one with a learning rate of 0.001. Therefore, this CNN model must be evaluated using the test dataset. In addition to the test accuracy, CNN evaluation

Table 4 Hyperparameters used in CNN training

Parameter	Value
Optimizer	Momentum [22]
Batch size	16
Learning rate	0.1, 0.01, and 0.001

currently includes the following measures: sensitivity (recall), specificity, positive predictive value (PPV, or precision), negative predictive value (NPV), and the area under the curve (AUC) for the receiver operating characteristic (ROC) curve, to evaluate the model's performance considering the data in the test dataset. The model's evaluation for each class based on the test data resulted in the values shown in Table 5. The confusion matrices in Table 6 summarize the overall and the specific results for each class. Another essential measure considered in this evaluation is the ROC curve. Figure 17 shows the curves for each class.

Observe that there is some disparity in the performance considering the three different classes, which is perceptible in confusion matrices (Table 6), the main test results (Table 5), and the ROC curves (Fig. 17). Nevertheless, the results suggest the applicability of CNNs for the proposed task.

Table 5 Test results

Class	Precision	Recall	Specificity	NPV	AUC-ROC
Normal	0.818	0.600	0.933	0.823	0.643
Incipient	0.722	0.866	0.833	0.926	0.861
Advanced	0.687	0.733	0.833	0.862	0.810

Table 6 Confusion matrix

		Predicted		
		Normal	Incipient	Advanced
True	Normal	60% (9)	13% (2)	27% (4)
	Incipient	7% (1)	86% (13)	7% (1)
	Advanced	7% (1)	20% (3)	73% (11)

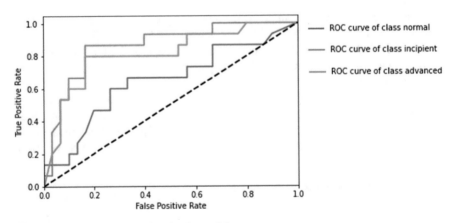

Fig. 17 ROC curves of each class for the model

3.4 Image Enhancement

The limitations in radiographic acquisition devices can result in low-resolution images, compromising the diagnosis process [43, 44]. Traditional image processing techniques, as interpolation methods, can be used to increase images' resolution. However, their results can be improved by AI-based methods.

Radiographic image enhancement also includes noise removal and image reconstruction, i.e., recovering missing parts of the image. Actually, for oral radiographs, these tasks are closely related since the missing data is comprehended as noise in this context. Moreover, noise removal demands reconstruction to replace this noisy data with the actual data. AI-based solutions are also popular for these tasks. As discussed in the section Challenge Issues, radiographs' acquisition and image formation processes can lead to a wide range of artifacts and noise. According to Schulze et al. [45], artifacts and noise in oral radiographs include blur, scatter artifacts, extinction artifacts (missing value), beam hardening artifacts, exponential edge gradient effects, aliasing artifacts, ring artifacts, and motion and misalignment artifacts. Another critical artifact that significantly affects the image quality is the metal artifact. Image processing techniques can aid the reduction of some of these artifacts, especially those that include AI-based techniques. There are a significant number of works in literature focused on artifact removal in dental imaging. Among them are the works of Wang et al. and Chang et al. [46, 47] that propose using neural networks for ring artifact removal in CBCT images. Xie et al. [48] present an algorithm based on convolutional neural networks to reduce scatter artifacts in CBCT. Zhang et al. [49] developed a convolutional neural network-based framework to reduce the effects of metal artifacts.

Increasing the Quality of Digital Periapical Radiographs Using SRCNN This section's example demonstrates the application of a widely known deep learning algorithm, called super-resolution convolution neural network (SRCNN) [50], to obtain high-resolution periapical images from low-resolution ones, reaching a magnitude improvement of 4x. Its results are compared with other super-resolution solutions based on more traditional image processing techniques, which are the nearest, bilinear, bicubic, and Lanczos interpolations.

SRCNN is a widely used deep learning-based super-resolution method. Dong [50] initially proposed it in 2016. In its pre-processing, the original low-resolution image is rescaled to its final size by applying the bicubic interpolation. Such a rescaled image is the input of the network that manipulated it in three main steps: patch extraction and representation, nonlinear mapping, and reconstruction. In the first step, patches are extracted from the bicubic rescaled image. Such patches are represented as high-dimensional vectors. In the second step, these high-dimensional vectors are mapped into other vectors, in a nonlinear way. In the third step, it aggregates the high-resolution patch-wise representations to obtain the output (high-resolution image). Figure 18 shows a representation of the steps that compose the SRCNN.

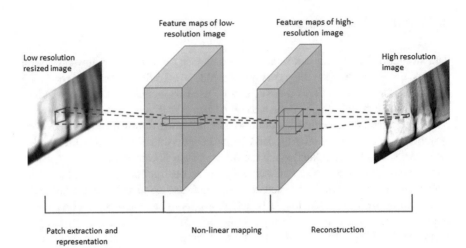

Fig. 18 SRCNN representation

Table 7 Evaluative measures for each considered method

Method	MSE	PSNR	SSIM
Nearest interpolation	101.04 (±47.35)	28.79 (±3.08)	0.92 (±0.02)
Bilinear interpolation	62.06 (±29.89)	31.16 (±3.48)	0.94 (±0.03)
Bicubic interpolation	53.39 (±26.01)	31.87 (±3.63)	0.94 (±0.02)
Lanczos interpolation	48.67 (±24.07)	32.33 (±3.75)	0.95 (±0.02)
SRCNN	33.69 (±14.17)	33.53 (±2.86)	0.98 (±0.01)

The training process of the SRCNN model included 10,000 epochs and used the Adam optimizer and a learning rate of 3×10^{-4}. The dataset used in the training process of the SRCNN model is composed of 228 different periapical radiographs.

After the training process, the obtained model must be evaluated considering the test set that refers to a new set of images not used for training. The test set is formed by 100 selected periapical radiographs, from the 120 that compose the original dataset provided by Rad et al. [51]. Such radiographs were collected in the Dental Clinic of the Universiti Teknologi Malaysia (UTM) Health Center using a Sirona device. All images are in grayscale, in "jpeg" format, with dimensions of 748×512. Also, a padding operation is applied to the images of both training and testing sets, in order to obtain squared images, with the same number of rows and columns, in order to facilitate the SRCNN processing.

The analysis of the results included three metrics to evaluate the similarity between the images achieved by the considered methods and the ground-truth high-resolution images: mean square error (MSE), peak signal-to-noise ratio (PSNR), and structural similarity index measure (SSIM). The results for all considered methods using the test dataset are presented in Table 7.

Note that the values presented in Table 7 demonstrate the SRCNN model's superiority, which outperformed all other methods for all considered measures. This

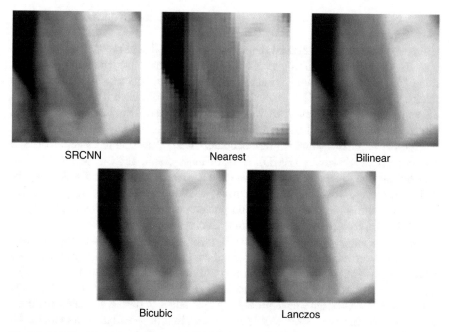

SRCNN   Nearest   Bilinear

Bicubic   Lanczos

Fig. 19 Detail of images generated by the considered methods

quality increase can also be observed by visually analyzing the images generated by each method (Fig. 19). Note the aliasing effects of the nearest interpolation's image and the blur effects of the bilinear's, bicubic's, and Lanczos' images. On the other hand, the SRCNN model led to less noise and more detailed edges.

3.5 Other Applications

There is a wide range of dental imaging applications for which the techniques covered in this chapter can be applied, as mentioned in the Introduction of this chapter. Both Schwendicke et al. [2] and Hung et al. [4] pointed out the localization of cephalometric landmarks is a very popular application. Cephalometric landmark localization in dental practice is performed manually by experts or supported by computerized tools, mostly in a semi-automatic way. As an attempt to automatize this process, several AI-based image processing solutions have been proposed in the last years. As discussed in the Challenge Issues section of this chapter, there is no consensus on the number of landmarks to be used in dental practice, so the works in literature that cover this topic vary by considering from 10 to 43 landmarks. Half of the works analyzed by Hung et al. [4] presented results considered by the authors as promising, with accuracy values ranging from 35% to 84.70%. Nevertheless, the quality of the results of the solutions available in literature still does not suit clinical requirements.

In that context, Arik [52] used shape-based CNN models to recognize landmarks' appearance patterns, defining probabilistic estimations for landmark locations. Song [53] proposed a two-step automatic method to detect cephalometric landmarks, which consists of (1) extracting image patches for each landmark and (2) – detecting the associated landmark in each patch using a ResNet model. The network directly outputs the coordinates of the landmarks.

Another popular application in this context is the detection of osteoporosis and low bone mineral density (BMD). Both of these conditions can be identified in radiographs due to their radiodensity-related aspects. Recent works for these applications achieved around 95% for accuracy, sensitivity, and specificity, suggesting that their inclusion in real-world dental practice is close. A great amount of these works defined features to be used as input in classifiers [54–57].

Diagnosis and segmentation of maxillofacial cysts and tumors using the mentioned tools are also commonly assessed in literature. The work presented by Abdolali et al. [58] considers the symmetry of oral anatomy to identify areas referent to cysts. Mikulka et al. and Nurtanio et al. [59, 60] proposed semi-automatic solutions using AI-based image processing techniques to detect, segment, and classify lesions of this type. More recently, Lee et al. [61] proposed using a GoogLeNet Inception v3 model to detect and classify odontogenic keratocysts, dentigerous cysts, and periapical cysts in CBCT, achieving an AUC of 0.914, a sensitivity of 96.1%, and a specificity of 77.1%. Kwon et al. [62] developed a CNN model inspired by the YOLOv3 architecture to detect and classify odontogenic cysts and tumors, which present 88.9% for sensitivity, 97.2% for specificity, 95.6% for accuracy, and 0.94 for AUC.

Other application areas in dental imaging are detection, segmentation, and classification of other anatomical structures, including teeth, jaw bone, and root canals; biofilm classification; diagnosis of multiple dental diseases; classification of tooth types; identification of inflamed gum; identification of dental plaque; and classification of the lower third molar stages.

3.6   Conclusions and Challenge Issues

Although there are several clinical decision support systems that have been developed in the last years, few of them are actually used in clinical settings and previous studies denote a low clinical acceptance of them [63–65], even considering that there is a consensus about their improvement in care and promotion on experts' efficiency [66]. That is, the great potential of the computational tools for dental image analysis based on image processing and artificial intelligence techniques still faces a significant amount of resistance from dentists and oral radiologists. In part, this may happen due to their novelty aspect: Research works employing CNNs in dentistry started in 2015 [2, 3]. This also could be related to the fact that the great majority of AI-based solutions do not consider the dentist comprehension factor, working basically as a *black box*, which affects their reliability hugely from the users' point of view.

However, even the use of computer-aided image examinations is not always well received by dental experts, who tend to demand second opinions for these evaluations, if used, since they believe it leads to an inconclusive diagnosis [67]. It is observed that the low quality of the images is one of the reasons that hugely present influences in such resistance and difficult development of user-friendly tools, which can aid in computer-aided diagnostic popularization and more uses [67]. Consequently, for oral diseases, manual clinical evaluation is still the gold standard in diagnosis.

Other critical aspects related to the difficulties of the development of AI systems are the subjectivity in expert conclusions and the lack of standards for some oral disease diagnosis. The perception of caries severity may vary among experts; for instance, it is common that there is no agreement about the amounts of the teeth that must be compromised in order to consider a lesion as incipient or not. As observed by Dave [67], there is experts' judgment disagreement even for defining more concrete points when a patient presents an anatomic abnormality. This hugely affects the development of public databases with a ground truth that must be used for computational based solutions in the dental imaging area since there is a hidden feeling related to fear of peers' opinion about the correctness of the report made (*i.e.* based on the definition of the diagnosis), so the experts' annotations are considered by the physicians as a risk to them because it may not be considered correct by other experts and consequently it is almost impossible to find one or promote construction of databases. Moreover, this tends to restrict the applications that can be considered in the development of computer-based solutions, and this aspect could be reduced when there are widely accepted diagnosis standards as in the case of BIRADS degree for breast researches [68].

Truthfully, this lack of public available data is the main and critical challenge issue to be considered, faced, and resolved in dental imaging applications. Very few open datasets are available, with the "ISBI 2015 Grand Challenge in Dental X-ray Image Analysis" being the most popular one. Most works in literature use private datasets from their associated institutions, which can lead to bias since different institutions tend to target different populations [2, 3] and so their databases are more representative and trustworthy. For example, public emergency hospitals tend to attend more to vulnerable patients with more severe lesions and often more neglected oral health. A dataset acquired in private institutions may have a higher number of healthy patients, preventing demographically correct representations and promoting the construction of non-generic solutions due to sub-representation aspects.

Oral radiographs are also more susceptible to present artifacts since dental prostheses and implants are substantially more prevalent than other body regions. Artifacts greatly affect dental radiographic images preventing a quality diagnosis and influencing the signal patterns used in detection algorithms. Moreover, oral radiographs are also influenced by same phenomena that affect radiographs in general, as acquisition problems and noise resultant from limitations in the image formation. The radio densities of some oral structures are difficult to detect in several oral diseases [69]. For example, bitewing images present a low sensitivity for both proximal and occlusive surfaces, and oral radiographs, in general, have a poor performance for detecting noncavity lesions.

Finally, oral diseases are heterogeneous and hard to model computationally (even impracticable sometimes) [2, 3, 6, 27, 70], restricting the application problems in which the proposed methods can be applied [32, 36, 43, 56, 58, 69].

References

1. R. Abdalla-Aslan, T. Yeshua, D. Kabla, I. Leichter, C. Nadler, An artificial intelligence system using machine-learning for automatic detection and classification of dental restorations in panoramic radiography: Automated detection and classification of panoramic dental restoration. Oral Surg. Oral Med. Oral Pathol. Oral Radiol. **130**(5), 593–602 (2020)
2. F. Schwendicke, T. Golla, M. Dreher, J. Krois, Convolutional neural networks for dental image diagnostics: A scoping review. J. Dent. **91**, 103226 (2019)
3. F. Schwendicke, W. Samek, J. Krois, Artificial intelligence in dentistry: Chances and challenges. J. Dent. Res. **99**(7), 769–774 (2020)
4. K. Hung, C. Montalvao, R. Tanaka, T. Kawai, M.M. Bornstein, The use and performance of artificial intelligence applications in dental and maxillofacial radiology: A systematic review. Dentomaxillofacial Radiol. **49**, 20190107 (2020)
5. K. Najarian, R. Splinter, *Biomedical Signal and Image Processing* (Taylor and Francis, 2012)
6. S.C. White, M.J. Pharoah, *Oral Radiology: Principles and Interpretation* (Elsevier Health Sciences, 2014)
7. P. Sukovic, Cone beam computed tomography in craniofacial imaging. Orthodontics Craniofacial Res. **6**, 31–36 (2003)
8. M.P. Bartold, 2006 periodontal tissues in health and disease: Introduction. Periodontol. **40**, 7–10 (2000)
9. R. Farina, A. Simonelli, A. Rizzi, L. Trombelli, Effect of smoking status on pocket probing depth and bleeding on probing following non-surgical periodontal therapy. Minerva Stomatol. **59**, 1–12 (2010)
10. K.S. Kornman, Mapping the pathogenesis of periodontitis: A new look. J. Periodontol. **8**, 1560–1568 (2008)
11. P.L. Lin, P.W. Huang, P.Y. Huang, H.C. Hsu, Alveolar bone-loss area localization in periodontitis radiographs based on threshold segmentation with a hybrid feature fused of intensity and the H-value of fractional Brownian motion model. Comput. Methods Prog. Biomed. **121**, 117–126 (2015)
12. P.L. Lin, P.Y. Huang, P.W. Huang, Automatic methods for alveolar bone loss degree measurement in periodontitis periapical radiographs. Comput. Methods Prog. Biomed. **148**, 1–11 (2017)
13. J.H. Lee, D.H. Kim, S.N. Jeong, S.H. Choi, Diagnosis and prediction of periodontally compromised teeth using a deep learning-based convolutional neural network algorithm. J. Periodontal Implant Sci. **48**, 114–123 (2018)
14. D.P. Carmody, S.P. McGrath, S.M. Dunn, P.F. van der Stelt, E. Schouten, Machine classification of dental images with visual search. Acad. Radiol. **8**, 1239–1246 (2001)
15. A. Mol, P.F. van der Stelt, Application of computer-aided image interpretation to the diagnosis of periapical bone lesions. Dentomaxillofac. Radiol. **21**, 190–194 (1992)
16. T. Ekert, J. Krois, L. Meinhold, K. Elhennawy, R. Emara, T. Golla, F. Schwendicke, Deep learning for the radiographic detection of apical lesions. J. Endod. **45**, 917–922 (2019)

17. J. Krois, T. Ekert, L. Meinhold, T. Golla, B. Kharbot, A. Wittemeier, F. Schwendicke, Deep learning for the radiographic detection of periodontal bone loss. Sci. Rep. **9**, 1–6 (2019)

18. M.K. Jeffcoat, I.C. Wang, M.S. Reddy, Radiographic diagnosis in periodontics. Periodontol. **7**, 54–68 (1995)

19. S.M. Pizer, E.P. Amburn, J.D. Austin, R. Cromartie, A. Geselowitz, T. Greer, B.H. Romeny, J.B. Zimmerman, K. Zuiderveld, Adaptive histogram equalization and its variations. Comput. Vis. Graph. Image Process **39**, 355–368 (1987)

20. K. He, X. Zhang, S. Ren, J. Sun, Deep residual learning for image recognition, in *Proceedings of the IEEE Conference on Computer Vision and Pattern Recognition*, (2016), pp. 770–778. https://doi.org/10.1109/CVPR.2016.90

21. C. Szegedy, W. Liu, Y. Jia, P. Sermanet, S. Reed, D. Anguelov, E. Dumitru, V. Vanhoucke, A. Rabinovich, Going deeper with convolutions, in *Proceedings of the IEEE conference on computer vision and pattern recognition*, (2015), pp. 1–9. https://doi.org/10.1109/CVPR.2015.7298594

22. I. Sutskever, J. Martens, G. Dahl, G. Hinton, On the importance of initialization and momentum in deep learning, in *International Conference on Machine Learning*. http://dl.acm.org/citation.cfm?id=3042817.3043064, (2013), pp. 1139–1147

23. Y. Bengio, Rmsprop and equilibrated adaptive learning rates for nonconvex optimization. CoRR abs/1502.04390 (2015)

24. T. Chai, R.R. Draxler, Root mean square error (RMSE) or mean absolute error (MAE)?–Arguments against avoiding RMSE in the literature. Geosci. Model Dev. **7**, 1247–1250 (2014)

25. O. Russakovsky, J. Deng, H. Su, J. Krause, S. Satheesh, S. Ma, Z. Huang, A. Karpathy, A. Khosla, M. Bernstein, Imagenet large scale visual recognition challenge. Int. J. Comput. Vision **115**, 211–252 (2015)

26. D. Powers, Evaluation: From precision, recall and f-measure to roc, informedness, markedness and correlation. J. Mach. Learn. Technol. **2**, 37–63 (2007)

27. C. Anderson, M. Curzon, C. Van Loveren, C. Tatsi, M. Duggal, Sucrose and dental caries: A review of the evidence. Obes. Rev. **10**, 41–54 (2009)

28. M. Balakrishnan, R.S. Simmonds, J.R. Tagg, Dental caries is a preventable infectious disease. Aust. Dent. J. **45**, 235–245 (2000)

29. P.W. Caufield, A.L. Griffen, Dental caries: An infectious and transmissible disease. Pediatr. Clin. N. Am. **47**, 1001–1019 (2000)

30. W.J. Loesche, Role of streptococcus mutans in human dental decay. Microbiol. Rev. **50**, 353 (1986)

31. P.D. Marsh, Microbiology of dental plaque biofilms and their role in oral health and caries. Dent. Clin. **54**, 441–454 (2010)

32. A. Wenzel, H. Hintze, Perception of image quality in direct digital radiography after application of various image treatment filters for detectability of dental disease. Dentomaxillofacial Radiol. **22**, 131–134 (1993)

33. Z. Akarslan, M. Akdevelioglu, K. Gungor, H. Erten, A comparison of the diagnostic accuracy of bitewing, periapical, unfiltered and filtered digital panoramic images for approximal caries detection in posterior teeth. Dentomaxillofacial Radiol. **37**, 458–463 (2008)

34. S.B. Dove, W. McDavid, A comparison of conventional intra-oral radiography and computer imaging techniques for the detection of proximal surface dental caries. Dentomaxillofacial Radiol. **21**, 127–134 (1992)

35. M.Z. Booshehry, A. Davari, F.E. Ardakani, M.R.R. Nejad, Efficacy of application of pseudo-color filters in the detection of interproximal caries. J. Dent. Res. Dent. Clin. Dent. Prospects **4**, 79 (2010)

36. N. Karimian, H.S. Salehi, M. Mahdian, H. Alnajjar, et al., A deep learning classifier with optical coherence tomography images for early dental caries detection, in *Proceedings of SPIE 10473: Lasers in Dentistry XXIV*, (2018), pp. 10473041–10473048

37. J.H. Lee, D.H. Kim, S.N. Jeong, S.H. Choi, Detection and diagnosis of dental caries using a deep learning-based convolutional neural network algorithm. J. Dent. **77**, 106–111 (2018)

38. M. Sornam, M. Prabhakaran, Logit-based artificial bee colony optimization (lb-abc) approach for dental caries classification using a back propagation neural network, in *Integrated Intelligent Computing, Communication and Security*, (2019), pp. 79–91
39. X. Xu, S. Xu, L. Jin, E. Song, Characteristic analysis of otsu threshold and its applications. Pattern Recogn. Lett. **32**, 956–961 (2011)
40. J. Serra, Morphological filtering: An overview. Signal Process. **38**, 3–11 (1994)
41. J. Schmidhuber, Deep learning in neural networks: An overview. Neural Netw. **61**, 85–117 (2015)
42. G. Litjens, T. Kooi, B.E. Bejnordi, A.A.A. Setio, F. Ciompi, M. Ghafoorian, C.I. Sanchez, A survey on deep learning in medical image analysis. Med. Image Anal. **42**, 60–88 (2017)
43. S. Kositbowornchai, M. Basiw, Y. Promwang, H. Moragorn, N. Sooksuntisakoonchai, Accuracy of diagnosing occlusal caries using enhanced digital images. Dentomaxillofacial Radiol. **33**, 236–240 (2004)
44. J. De Morais, C.E. Sakakura, L.C.M. Loffredo, G. Scaf, Accuracy of zoomed digital image in the detection of periodontal bone defect: In vitro study. Dentomaxillofacial Radiol. **35**, 139–142 (2006)
45. R. Schulze, U. Heil, D. GroB, D.D. Bruellmann, E. Dranischnikow, U. Schwanecke, E. Schoemer, Artefacts in CBCT: A review. Dentomaxillofacial Radiol. **40**, 265–273 (2011)
46. Z. Wang, L. Jianwu, E. Mogendi, Removing ring artifacts in CBCT images via generative adversarial networks with unidirectional relative total variation loss. Neural Comput. & Applic. **31**, 5147–5158 (2019)
47. S. Chang, X. Chen, J. Duan, X. Mou, A CNN based hybrid ring artifact reduction algorithm for CT images. IEEE Trans. Radiat. Plasma Med. Sci. **5**(2), 253–260 (2020)
48. S. Xie, C. Yang, Z. Zhang, H. Li, Scatter artifacts removal using learning-based method for CBCT in IGRT system. IEEE Access **6**, 78031–78037 (2018)
49. Y. Zhang, Y. Hengyong, Convolutional neural network based metal artifact reduction in x-ray computed tomography. IEEE Trans. Med. Imaging **37**, 1370–1381 (2018)
50. C. Dong, C.C. Loy, K. He, X. Tang, Image super-resolution using deep convolutional networks. IEEE Trans. Pattern Anal. Mach. Intell. **38**, 295–307 (2015)
51. A.E. Rad, M.S.M. Rahim, A. Rehman, R. Saba, Digital dental X-ray database for caries screening. 3D. Research **7**, 18 (2016)
52. S.O. Arik, B.U. Ibragimov, L. Xing, Fully automated quantitative cephalometry using convolutional neural networks. J. Med. Imaging **4**, 014501 (2017)
53. Y. Song, X. Qiao, Y. Iwamoto, Y.W. Chen, Automatic cephalometric landmark detection on X-ray images using a deep-learning method. Appl. Sci. **10**, 2547 (2020)
54. J.J. Hwang, J.-H. Lee, S.-S. Han, Y.H. Kim, H.-G. Jeong, Y.J. Choi, Strut analysis for osteoporosis detection model using dental panoramic radiography. Dentomaxillofac. Radiol. **46**, 20170006 (2017)
55. M.S. Kavitha, A. Asano, A. Taguchi, M.-S. Heo, The combination of a histogram-based clustering algorithm and support vector machine for the diagnosis of osteoporosis. Imaging Sci. Dent. **43**, 153–161 (2013)
56. M.S. Kavitha, A. Asano, A. Taguchi, T. Kurita, M. Sanada, Diagnosis of osteoporosis from dental panoramic radiographs using the support vector machine method in a computer-aided system. BMC Med. Imaging **12**, 1 (2012)
57. M.S. Kavitha, P. Ganesh Kumar, S.-Y. Park, K.-H. Huh, M.-S. Heo, T. Kurita, Automatic detection of osteoporosis based on hybrid genetic swarm fuzzy classifier approaches. Dentomaxillofac. Radiol. **45**, 20160076 (2016)
58. F. Abdolali, R.A. Zoroofi, Y. Otake, Y. Sato, Automatic segmentation of maxillofacial cysts in cone beam CT images. Comput. Biol. Med. **72**, 108–119 (2016)
59. J. Mikulka, E. Gescheidtova, M. Kabrda, V. Perina, Classification of jaw bone cysts and necrosis via the processing of Orthopantomograms. Radioengineering **22**, 114–122 (2013)

60. I. Nurtanio, E.R. Astuti, I. Ketut Eddy Pumama, M. Hariadi, M.H. Purnomo, Classifying cyst and tumor lesion using support vector machine based on dental panoramic images texture features. IAENG Int. J. Comput. Sci. **40**, 29–37 (2013)

61. J.H. Lee, D.H. Kim, S.N. Jeong, Diagnosis of cystic lesions using panoramic and cone beam computed tomographic images based on deep learning neural network. Oral Dis. **26**, 152–158 (2020)

62. O. Kwon, T.H. Yong, S.R. Kang, J.E. Kim, Automatic diagnosis for cysts and tumors of both jaws on panoramic radiographs using a deep convolution neural network. Dentomaxillofacial Radiol. **49**, 20200185 (2020)

63. B. Kaplan, Evaluating informatics applications—Clinical decision support systems literature review. Int. J. Med. Inf. **64**, 15–37 (2001)

64. J.G. Anderson, J.J. Stephen, *Use and Impact of Computers in Clinical Medicine* (Springer Verlag, New York, 1987)

65. B. Kaplan, Evaluating informatics applications—Some alternative approaches: Theory, social interactionism, and call for methodological pluralism. Int. J. Med. Inf. **64**, 39–56 (2001)

66. E.A. Balas, S.A. Boren, Clinical trials and information interventions, in *Clinical Decision Support Systems: Theory and Practice*, (Springer, New York, 1999), pp. 199–216

67. M. Dave, K. Horner, Challenges in X-ray diagnosis: A review of referrals for specialist opinion. Br. Dent. J. **222**, 431–437 (2017)

68. A. Conci, A. Sanchez, P. Liatsis, H. Usuki, Signal processing techniques for detection of breast diseases. Signal Process. **93**, 2783–2784 (2013)

69. A. Goren, Current and future trends in dental radiography. Oral Health Dent. Manage. **14**, 41 (2015)

70. A. Yaji, S. Prasad, A. Pai, Artificial intelligence in dento-maxillofacial radiology. Acta Sci. Dent. Sci. **3**, 116–121 (2019)

Median Filter Based on the Entropy of the Color Components of RGB Images

José Luis Vázquez Noguera ⓘ, Horacio Legal-Ayala ⓘ,
Julio César Mello Román ⓘ, Derlis Argüello, and Thelma Balbuena

1 Introduction

Nonlinear order filters are a tool used for the removal of noise in digital images. The main advantage of these filters is that they preserve the characteristic structures of the images, such as the edges [3]. In contrast, linear filters apply a blur to all image structures [3]. The main challenge of the filters is to have the ability to differentiate the desired content from the noise. Addressing this problem, many order filters have been designed [1, 5–7, 11, 12, 21].

The element that is selected in each filter window depends on the type of the order filter chosen. For example, a filter that has been shown to efficiently remove noise from grayscale images replaces each pixel value with the median value of the pixel's neighborhood. This type of filter is known as a median filter.

Initially, order filters were designed for grayscale images. These filters have trivial implementations because the intensities are scalar values and ordering them has no major complications. Currently, research is focusing its energy on the extension of grayscale algorithms to color images [10, 11, 14, 15, 21]. This is because a color image provides more information about the captured scene. Applying order filters that directly process pixel colors, without prior conversion to grayscale, is a challenge today. A universal method has not yet been adopted, but there are already different implementations with advantages and disadvantages.

Noise removal is an image pre-processing step. Noise can hinder specific processing, such as object recognition and segmentation. In medical imaging there is a need to reduce noise [22]. For example, a common problem in ultrasound imaging is speckle noise, which is caused by errors in data transmission. Another known noise is the Gaussian noise. This usually appears during the image acquisition stage

J. L. V. Noguera (✉) · H. Legal-Ayala · J. C. M. Román · D. Argüello · T. Balbuena
Facultad Politécnica, Universidad Nacional de Asunción, San Lorenzo, Paraguay
e-mail: jlvazquez@pol.una.py; hlegal@pol.una.py; juliomello@pol.una.py

© Springer Nature Switzerland AG 2022
P. Johri et al. (eds.), *Trends and Advancements of Image Processing and its Applications*, EAI/Springer Innovations in Communication and Computing, https://doi.org/10.1007/978-3-030-75945-2_5

[4]. Also, a Gaussian noise is usually present in the transmission step, because of random fluctuations in the signal [13].

The present proposal aims to eliminate noise in digital color images. The noise is removed by an order filter based on the entropy of the color components in RGB images. Specifically, the objective of the proposal is to remove Gaussian and speckle noise with greater efficiency than the state-of-the-art proposals, using as a metric the mean absolute error (MAE).

The article is organized in five sections. Section 2 presents the main vectorial ordering strategies found in the state of the art. Section 3 presents the development of the proposal. Section 4 presents the results obtained in the experiments. Finally, Section 5 presents the conclusions and future work.

2   Ordering Strategies

An image is an $f: Z^2 \rightarrow Z^n$ function. Each $(x, y) \in Z^2$ pair is one pixel, and $f(x, y) \in Z^n$ is the color of the image at the (x, y) pixel. For k bits (commonly 8), $f(x, y) = (C_1, C_2, C_3)$, and $C_1 \in \{0, 1, \cdots, 2^k - 1\}$ is the intensity of the red component, $C_2 \in \{0, 1, \cdots, 2^k - 1\}$ is the intensity of the green component, $C_3 \in \{0, 1, \cdots, 2^k - 1\}$ is the intensity of the blue component, and $f(x, y)$ is the color resulting from mixing these components in the pixel (x, y).

The *component histogram* function is defined from the RGB image, which corresponds to the frequency distribution of the values that an image can take in one of the components. The histogram of the i-th component of the color image f is a discrete function $h_{f_i}^D$ and is defined as:

$$h_{f_i}^D (j) = n_j, \tag{1}$$

where j represents an intensity level in the range $\{0, 1, \cdots, 2^{k-1}\}$ of the i component of the image , and n_j is the number of pixels in the image whose intensity level is j in the D domain. In Fig. 1 we can see a color image without noise.

2.1   Ordering Techniques

The different vector ordering strategies can be classified according to the technique they use, that is, the criteria that are used to establish the order between the different elements of the set to be ordered.

According to [2], the vector ordering techniques are the following:

- Marginal ordering (M-ordering): The marginal ordering technique compares each color component independently. The main problem with this technique is that it generates false colors.

Fig. 1 Color image without noise

- Conditional ordering (C-ordering): Vectors are ordered by some marginal component selected sequentially according to different conditions.
- Partial ordering (P-ordering): The technique is based on the partitioning of the vectors into equivalence groups, such that there is an order between the groups. In this case, *partial* is an abuse of terminology, since there is total ordering that belongs to this particular class.
- Reduced ordering (R-ordering): Vectors are first reduced to scalar values and then classified according to their natural scalar order. For example, an R-ordering in Z^n may consist of first defining a transformation in $T : Z^n \rightarrow R$ and then ordering the colors with respect to the scalar order of their projection in Z^n by T.

2.2 Vector Ordering Strategies

The main vector ordering strategies, present in the state of the art, will be described below.

- Lexicographical ordering (LEX): The lexicographical ordering is a well-known example of conditional order. This order potentially uses all the available components of the vector. The disadvantage of this order is that it introduces arbitrariness into the prioritization of each color component [1].
- α-Lexicographical ordering (ALEX): The α-lexicographical ordering is a variation of the lexicographical ordering in which the aim is to reduce the number of times the order is decided by the first component by adding a α value to the first component of one of the vectors [12].

- α-Modulus lexicographical ordering (AMLEX): This order seeks to reduce the number of times, in which the order is decided by the first component. Also, it proposes the reduction of decisions by the first component. This is done by the division between the first component of both vectors and a constant value α [1].
- HSI lexicographical ordering (HLEX): In this lexicographical ordering strategy, vectors are expressed as coordinates of the HSI space. The use of this ordering strategy requires a parameter called *reference hue*, which determines what is the least color hue (this is necessary because of the circular nature of the hue component) [12].
- Euclidean distance (ED): This strategy uses the reduction technique. This defines the transformation value T of each color C as the Euclidean distance between C and the coordinate origin [12].
- Euclidean distance in the L*a*b* space (DLAB): This order strategy is similar to the previous one, with the difference being that the colors (vectors) are expressed as points in the L*a*b* color space [12].
- Bit mixing (BM): The bit mixing strategy uses the reduction technique, defining the transformation of each vector as the integer value obtained by mixing the bits that represent the intensities of each component of the vector. This reduction ordering strategy defines a total order in the RGB color set [5].
- Average as RGB component weight (MEAN): This is a reduction ordering strategy that depends on the input images, and defines the T transformation of each vector as a weighted sum of its elements [21], as follows:

$$T(C) = w_1 \times C_1 + w_2 \times C_2 + w_3 \times C_3, \tag{2}$$

where w_1, w_2, and w_3 are the weights of the R, G, and B components, respectively. Each weight w_i is defined according to this strategy as the average of the intensity values in the i component.

- Minimum as RGB component weight (MIN): This strategy uses the same concept as the previous strategy with the difference being that the weights are defined as the minimum intensity value of each component [11].
- Maximum as RGB component weight (MAX): This ordering strategy is analogue to the previous one; it defines the transformation as the maximum current value of each component [11].
- Mode as RGB component weight (MO): This ordering strategy defines the weight as the most repeated value (mode) of each component. In one domain it can be the case that there is more than one mode, so this strategy can be divided into two: one that takes the lower value and another that selects the higher value [11].
- Variance as RGB component weight (VAR): This strategy differs from the previous one by the weight used, since it calculates the variance of the set of values of each RGB component within a domain [11].

- Smoothness as RGB component weight (SMO): This is an ordering strategy like the previous one, with the difference being that the weight of each component is defined by a smoothness formula [11], as follows:

$$1 - \frac{1}{1 + \sigma_i^2},\tag{3}$$

where σ_i^2 is the variance of the i component.

3 Proposed Ordering Strategy

This section presents the proposal. First, the term entropy will be introduced, which is used as statistical information of the image and is a key piece of the proposed order.

3.1 Entropy

Entropy is a term used in thermodynamics as a measure of dispersion and disorder [19]. In information theory it is the quantification of the information provided by a set of values or symbols [18].

The term information entropy or Shannon entropy, introduced in 1943 by the mathematician and electronic engineer Claude Shannon, is a quantitative measure of information. With this measure, the problem of how to transmit information as efficiently as possible through a given physical channel was addressed [23]. Information entropy is defined as the amount of information gained when the result of a random process is inferred [18]. This is calculated as follows:

$$E(U) = - \sum_{u \in U} p(u) \log_2 p(u),\tag{4}$$

where $u \in U$ is a random variable and $p(u)$ is the probability of occurrence of the u value. This concept is widely used in different types of applications, as a process or as a measure to evaluate a result [16, 17, 20].

A set of symbols in which the probability of occurrence of each symbol is the same (uniformly distributed set) has higher entropy than a set in which the symbols have different probabilities of occurrence. That is, the information that is "gained" by predicting a symbol from a uniformly distributed set is more valuable because that symbol is more difficult to predict.

When the set of symbols from which entropy is to be calculated is a color component f_i of an image f in a domain D, this can be expressed to entropy as a function of the histogram of the component by calculating the probability of occurrence

$p_{f_i}^D(j)$ of each intensity level j as its value in the histogram divided by the total number of pixels within the domain.

$$p_{f_i}^D(j) = \frac{h_{f_i}^D(j)}{n(D)}, \tag{5}$$

where $h_{f_i}^D(j)$ is the histogram value of f_i for the intensity j and $n(D)$ is the cardinality of the D domain, i.e., the total number of pixels within the D domain.

The entropy of f_i in the D domain is defined as follows:

$$E_{f_i}^D = -\sum_{j=0}^{2^k-1} p_{f_i}^D(j)\log_2 p_{f_i}^D(j). \tag{6}$$

The concept of entropy was used in this work to assign a level of importance to each RGB color component within the image. Later, we will see the order used by the proposed filter. This filter has a transformation step from colors to scalar values, for which you need to assign weights to each color component. The weights are defined according to the entropy of the color components. Therefore, the importance that this ordering strategy assigns to each color component is based on the amount of information that the component gives, compared to the others.

3.2   Proposed Filter

The proposal of this work is a median filter that uses entropy information of the RGB color components. In the proposed filter, a lexicographic order is used. In this strategy a new value is in the first position of the lexicographic cascade. This is done so as not to give higher priority to a specific component of the vector that represents the color. The new calculated value corresponds to a transformation obtained from metrics associated with the histogram of each component (R, G, B) of the image.

RGB colors are reduced to a scalar value. Therefore, a $T : Z^3 \rightarrow R$ transformation is first defined and then the colors are sorted with respect to the scalar order of their projection in Z^3 by T. The reduction of a $C = (C_1, C_2, C_3)$ color is achieved by the internal product of RGB color with a $w = (w_1, w_2, w_3)$ weight vector, as follows:

$$T(C) = \sum_{i=1}^{3}(w_i \times C_i), \tag{7}$$

where i is the index of the component and $w_i \in R$.

Two colors, C and C', with $C \neq C'$, can have the same transformation, i.e., $T(C) = T(C')$. Therefore, the transformation is used as the first component of the lexicographical order.

The values of the vector w are obtained by applying a function $g \in R$ to the histogram of each component in a D domain of the image f that is $w_1 = g\left(h_{f_1}^D\right)$, $w_2 = g\left(h_{f_2}^D\right)$, and $w_3 = g\left(h_{f_3}^D\right)$, with $f_1 =$ component R, $f_2=$ component G, and $f_3 =$ component B.

The function g is obtained by applying the entropy metric to the histogram, in order to give more weight to that component whose entropy has a higher value in a specific D domain (in all or part of the image). In the experimental section, different domain distributions are presented in the application of the filter.

The entropy of a color component of an image within a domain was defined in Eq. 5, and the calculation of the weights w_i in a D domain according to entropy is as follows:

$$w_i = -\sum_{j=0}^{2^k-1} p_{f_i}^D(j)\log_2 p_{f_i}^D(j), \tag{8}$$

where i is the component identifier of the image f, $2^k - 1$ is the highest value of each component, and $p_{f_i}^D(j)$ is the probability of appearance of the intensity j in the component f_i and domain D.

With the development of w_i, the transformation function from Eq. 7 is as follows:

$$T(C) = \sum_{i=1}^{3}\left(\left(-\sum_{j=0}^{2^k-1} p_{f_i}^D(j)\log_2 p_{f_i}^D(j)\right)\times C_i\right). \tag{9}$$

In the case of equality in the first instance of the lexicographic cascade, we proceed to compare the value of the components C_i in the order C_1, C_2, C_3.

In order to formally define the proposed order between two colors C and C', the following expression is presented:

$$C < C' \leftrightarrow \left[T(C),C_1,C_2,C_3\right]^T <_L \left[T(C'),C_1',C_2',C_3'\right]^T, \tag{10}$$

where $<_L$ is the lexicographical ordering.

The following section presents the definition of the metric used to compare the different ordering strategies with the median filter. Also, the implementation of the proposal formulated for the realization of the experiments, as well as the results obtained by those experiments, is detailed.

4   Experiments and Results

This section describes the metrics that are used for the evaluation of the proposed filter in comparison with the studied state-of-the-art filters. It also describes the different parameters used for the implementation of the filters.

4.1 Mean Absolute Error (MAE)

MAE is a metric used in statistics to measure how accurate predictions are of actual outcomes [8]. Let us say an original RGB image f with dimensions $M \times N$ and its corresponding filtered image g, the MAE of the filtered image is given by:

$$d = \sum_{\substack{x \in \{1,2,\cdots,M\} \\ y \in \{1,2,\cdots,N\}}} \left| f(x,y) - g(x,y) \right|, \tag{11}$$

$$MAE = \frac{1}{3MN} \sum_{i=1}^{3} d_i. \tag{12}$$

4.2 Implementation of the Proposal

The filter proposed in this work defines a reduction technique based on entropy as an ordering strategy and the median as the element selected in each filter mask. Other aspects related to the application of the proposed filter are left to the user as parameters to be defined.

One parameter to be defined is the domain, which is considered in the calculation of the weights. Different domain distributions were used to test the proposed filter, which are explained below.

Windows as a Domain This technique divides the image previously in windows, as shown in Fig. 2.

For the calculation of the weights within a filter mask $S_{(x, y)}$ centered on a pixel (x, y), the domain D will correspond to the window W_b in which the mask is located.

Figure 3 shows the following elements:

- Filter mask $S_{(x, y)}$ (sky colored): the filter mask centered on the pixel (x, y) (lavender colored).
- Domain D: area of the image considered for the calculation of component weights, textured with yellow stripes.

In this case the domain D is the window W_1, because it contains the filter mask $S_{(x, y)}$. It should be noted that the filter mask does not have to be of the same size as the windows, as it happens in this example.

The number of windows is a parameter provided by the user. In Sect. 4.4 we show which values of this parameter had the best results experimentally.

As in [21], the problem of when two adjacent pixels belong to different windows and therefore would have different weights is addressed. In Fig. 4 you can see that the mask $S_{(x, y)}$ contains two windows, W_1 and W_2.

Fig. 2 Windows W_1, W_2, W_3, and W_4

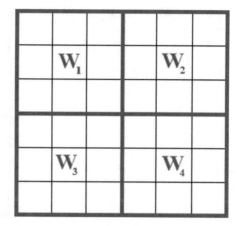

Fig. 3 Windows, mask, and domain

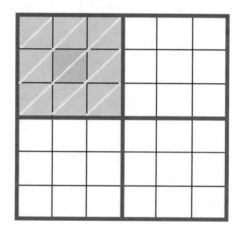

Fig. 4 Mask touching more than one window

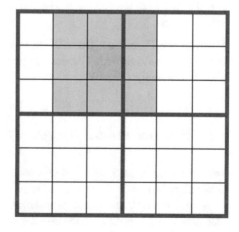

Fig. 5 Domain when the
mask touches more than
one window

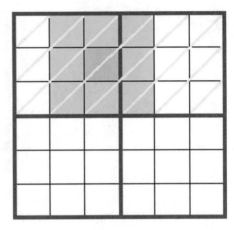

Fig. 6 Filter mask as a
domain

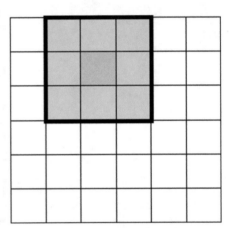

In this case the domain D becomes the set of windows W_i that have one or more
pixels of the filter mask $S_{(x,y)}$. In Fig. 5 the corresponding D domain is marked with
yellow stripes.

Mask of the Filter as a Domain In this technique, the step of previous division
into windows is omitted. The domain becomes each filter mask centered on each
pixel of the image (Fig. 6).

4.3 Experiments and Results

Tests were performed with 100 different images from a public database [9], con-
taminating them with Gaussian and speckle noise.

Note that a previous step of normalization of the input images to the range [0, 1]
was performed to generate the noises; then each image was scaled back to its original range [0, $2^k - 1$]. The parameters used in the generation of each type of noise are
described below.

Table 1 List of ordering strategies with their respective abbreviations

Ordering strategies	Abbreviation
Entropy (proposed filter)	ENT
Euclidean distance	ED
Bit mixing	BM
Lexicographic	LEX
α-lexicographical ordering	ALEX
α-modulus lexicographical ordering	AMLEX
HSI lexicographical ordering	HLEX
Euclidean distance in L*a*b*	DLAB
Minimum	MIN
Maximum	MAX
Mode 1	MO1
Mode 2	MO2
Smoothness	SMO
Mean	MEAN
Variance	VAR

To contaminate the images with Gaussian noise, we used mean $\mu = 0$ and varied the variance parameter σ^2 between values 0.005 and 0.165, making steps of 0.01. In the case of images with speckle noise, the parameters are also mean μ and variance σ^2. In the same way, the values used were $\mu = 0$ and σ^2 between the values 0.005 and 0.165, making steps of 0.01.

Table 1 shows the median filters that were subject to experimentation with the respective codes used to abbreviate their names.

There are two types of mode filters. The first one, called MO1, uses the lower value in case there is more than one mode. On the other hand, the method called MO2 uses the higher value.

For LEX, ALEX, and AMLEX all possible combinations of priority orders of RGB components were used.

The number of windows in which the images were divided (all equal in size) varied between 1, 9, 25, and 49. In the graphic representations of the filters, the suffix "W" and the number of windows are added to the filter abbreviation. For example, W9 implies that the image was divided into nine windows of the same size. The filter mask size was kept constant with a value of 3×3 for the filters with a priori window division, not for the implementations without window division, in which case it was varied between the sizes 3×3, 5×5, and 7×7. In the graphic representations of the filters, the suffix "M" and the mask size are added to the filter abbreviation. For example, M5 implies that a 5×5 mask is used for filtering.

The value of the ALEX and AMLEX parameter α was kept at a constant value of 10, which was optimal in the results of [1]. In HLEX the following reference hues were used: 0, 90, 180, and 270.

For the MAE metric used to evaluate the filters, low results translate into better results.

4.4 *Results*

This section shows the experimental results of the proposed filter compared to the state of the art.

Figure 7 shows an image (Fig. 7a) from the database [9] and the same image with Gaussian noise, with $\sigma^2 = 0.105$ and $\mu = 0$.

The result of applying the proposed order filter (ENT) and the other filters evaluated on the contaminated image can be seen in Fig. 8. ENT attenuates noise much better, preserving the most important original contours.

In the following subsections a summary of the numerical results obtained from the experiments will be provided. For each noise, the following types of graphs are provided:

- *Frequency bars*: This shows the number of images where each filter obtained the lowest average of metrics, i.e., the number of images for which each filter was better. *Note:* The number of input images is 100 but the sum of the values of a frequency bar graph can be greater than 100 due to "ties." This is because there may be different filters that are better for the same image. For filters using weights, the results will be summarized as a sum total of images for which that weight was better (without discriminating by domain configuration).
- *Trend curves*: It presents the average MAE obtained by each filter for a specific value of noise parameter σ^2 for Gaussian and speckle noise, that is, the points (σ^2, MAE). The curves are drawn to better visualize the trend shown by the filters as the noise parameter increases. The curve corresponding to a filter is obtained by joining each pair of successive points (σ^2, MAE) of that filter with the line (straight) passing through both points. Each trend curve will be accompanied by its respective table with the total sum of MAE obtained by each filter. For filters with domain configuration, only the results obtained from the best configuration will be shown.

Gaussian Noise The frequency bars corresponding to Gaussian noise are shown in Fig. 9. It can be seen that ENT is better for the largest number of images (62 out of 100).

The best filter, on average, for Gaussian noise is ENT with filter mask configuration of 5 × 5 as domain (ENTM5). It is remarkable that the 5 × 5 window configuration resulted to be the most effective for all filters with configurable domain. Figure 10 illustrates the situation described by showing a trend curve graph as the noise parameter increases, with a magnification at the last point.

Table 2 provides a summary of the curve graph shown, with the sum of the MAE values obtained at each point by each filter.

Speckle Noise Figure 11 corresponds to the frequency bar graph for speckle noise. ENT obtains the best results in a total of 76 images out of 100. In this figure you can see the marked difference between the filter with the proposed ordering and the others found in the state of the art.

(a) Original

(b) Gaussian noise

Fig. 7 Image taken from the database [7] and the original image with Gaussian noise ($\sigma^2 = 0.105$; $\mu = 0$). (**a**) Original. (**b**) Gaussian noise

Figure 12 is a trend curve graph as the noise parameter increases, with a magnification at the last point. The filter with the proposed ordering strategy always obtains lower MAE compared to the other state-of-the-art ordering strategies.

Table 3 shows the sum of the MAE values obtained by each filter in the set of σ^2 values used for the tests. The proposed filter has a lower sum of errors.

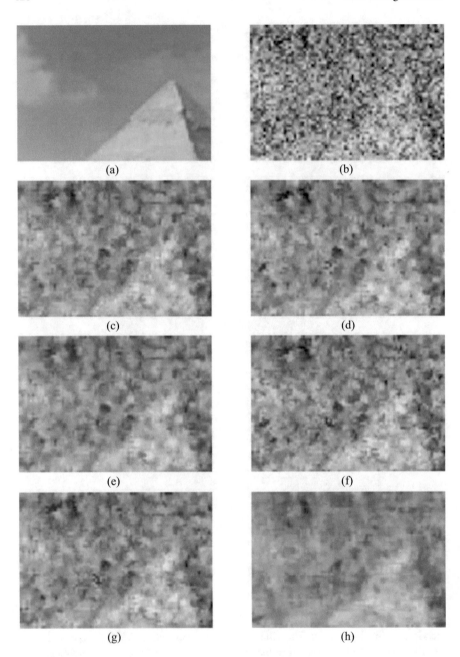

Fig. 8 The results of applying different filters evaluated on the image 7(b). The ENT filter uses the mask setting 5 × 5 as its domain. (**a**) Original. (**b**) Gaussian. (**c**) BM. (**d**) DLAB. (**e**) ED. (**f**) LEX. (**g**) HLEX. (**h**) ENT

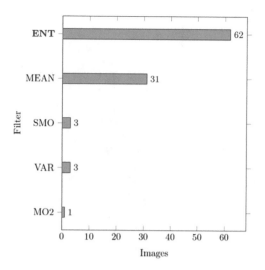

Fig. 9 Number of images where each filter obtained the best average. Gaussian noise. Summary by weight

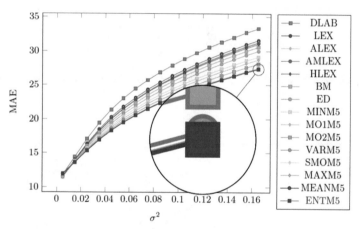

Fig. 10 Gaussian noise. MAE for σ^2. Best configurations

Table 2 Gaussian noise. Best configurations. Sum of MAE for σ^2

Filter	Sum
ENTM5	362.9703
MEANM5	363.0851
MAXM5	363.2554
SMOM5	363.3199
VARM5	364.6676
MO2M5	370.2034
MO1M5	376.4257
MINM5	381.5718
ED	389.1947
BM	396.2218
HLEX	402.5317
AMLEX	403.401
ALEX	404.8423
LEX	408.5205
DLAB	430.5146

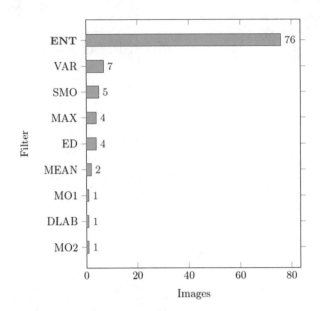

Fig. 11 Number of images where each filter obtained the best average. Speckle noise. Summary by weight

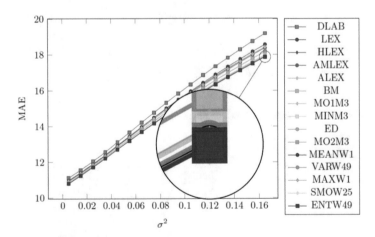

Fig. 12 Speckle noise. MAE for σ^2. Best configurations

Table 3 Speckle noise. Best configurations. Sum of MAE for σ^2

Filter	Sum
ENTW49	246.5442
SMOW25	246.5463
MAXW1	246.5472
VARW49	246.6322
MEANW1	246.6683
MO2M3	246.817
ED	246.9263
MINM3	247.2852
MO1M3	247.3627
BM	249.2121
ALEX	251.6364
AMLEX	251.687
HLEX	252.4788
LEX	253.0532
DLAB	258.8935

5 Conclusions and Future Work

A new RGB color ordering strategy is proposed, which is image dependent. Ordering is done by extracting entropy information from the histogram of each color component in a certain image domain and thus assigning weights to each of the components. These weights are used to transform a color to a scalar value using the reduction technique, which is evaluated in the first instance in the proposed order, and in case of equality in the reduction component, the lexicographic technique is used over the other components.

The proposed strategy establishes a median filter for the removal of speckle and Gaussian noise in digital color images in the RGB color space. Entropy was selected as a weight to perform the filter order because it is widely used in information theory as a quantitative measure of the information provided by a set of symbols. This way you can take advantage of the fact that a set of possible values of intensity in the color components of the images can be a set of symbols. In this way the concept of information entropy can be transferred to digital image processing.

To validate the proposal, a comparison was made of the proposed filter with various order filters found in the state of the art. This comparison was made by applying the filters to a sample of images contaminated with speckle and Gaussian noises, varying the noise parameters and using the MAE metric to measure the error. Tests showed that the proposed method has an average effectiveness of 62% for Gaussian noise removal, and for speckle noise, it performed even better, with 76% effectiveness compared to the other state-of-the-art ordering strategies.

References

1. E. Aptoula, S. Lefevre, On lexicographical ordering in multivariate mathematical morphology. Pattern Recogn. Lett. **29**(2), 109–118 (2008)
2. V. Barnett, The ordering of multivariate data. J. Royal Statist. Soc. Ser. A (General), **139**, 318–355 (1976)
3. W. Burger, M.J. Burge, *Digital Image Processing: An Algorithmic Introduction Using Java* (Springer Science & Business Media, 2009)
4. D.P. Cattin, Image restoration: Introduction to signal and image processing. MIAC, University of Basel. Retrieved on 11.11.2020 from https://docplayer.net/85966756-Image-restoration-introduction-to-signal-and-image-processing-prof-dr-philippe-cattin-miac-university-of-basel-april-19th-26th-2016.html
5. J. Chanussot, P. Lambert, Bit mixing paradigm for multivalued morphological filters. In: *IEE Conference Publication*, vol. 2, pp. 804–808. Institution of Electrical Engineers (1997)
6. M.L. Comer, E.J. Delp, Morphological operations for color image processing. J. Electr. Imag. **8**(3), 279–289 (1999)
7. A. Hanbury, J. Serra, Mathematical morphology in the cielab space. Image Anal. Stereol. **21**(3), 201–206 (2002)
8. R. Lukac, K.N. Plataniotis, B. Smolka, A.N. Venetsanopoulos, Generalized selection weighted vector filters. EURASIP J. Adv. Sig. Proc. **2004**(12), 1–16 (2004)
9. D. Martin, C. Fowlkes, D. Tal, J. Malik, A database of human segmented natural images and its application to evaluating segmentation algorithms and measuring ecological statistics. In: *Proceedings Eighth IEEE International Conference on Computer Vision. ICCV 2001. IEEE Comput. Soc.*
10. R. Mendez, R. Cardozo, J.L.V. Noguera, H. Legal-Ayala, J.C.M. Román, S. Grillo, M. García-Torres, Color image enhancement using a multiscale morphological approach. In: *Communications in Computer and Information Science*, pp. 109–123. Springer International Publishing (2019)
11. J.L.V. Noguera, C.E. Schaerer, J. Facon, H.L. Ayala, Adaptive RGB color lexicographical ordering framework using statistical parameters from the color component histogram. IEEE Access **7**, 141738–141753 (2019)
12. O. Zamora, F. Gabriel, Procesamiento morfológico de imágenes en color: aplicación a la reconstrucción geodésica. PhD Dissertation. Universidadde Alicante. http://hdl.handle.net/10045/10053 ISBN: 84-688-0091-0. (2002)
13. C.S. Panda, S. Patnaik, Filtering corrupted image and edge detection in restored grayscale image using derivative filters. Int. J. Image Proc. **3**(3), 105–119 (2009)
14. C. Riveros, H. Morel, H.L. Ayala, J.L.V. Noguera, Color ordering strategy based on loewner order applied to the mathematical morphology. In: *2016 XLII Latin American Computing Conference (CLEI)*. IEEE (Oct 2016)
15. J.C.M. Roman, H.L. Ayala, J.L.V. Noguera, Image color contrast enhancement using multiscale morphology. J. Comput. Interdiscip. Sci. **8**(3), 119–129. http://dx.doi.org/10.6062/jcis.2017.08.03.0129 (2017)
16. J.M. Román, J.V. Noguera, H. Legal-Ayala, D. Pinto-Roa, S. Gomez-Guerrero, M.G. Torres, Entropy and contrast enhancement of infrared thermal images using the multiscale top-hat transform. Entropy **21**(3), 244 (2019)
17. M.S. Roodposhti, J. Aryal, A. Lucieer, B. Bryan, Uncertainty assessment of hyperspectral image classification: Deep learning vs. random forest. Entropy **21**(1), 78 (2019)
18. C.E. Shannon, A mathematical theory of communication. ACM SIGMOBILE Mobile Computi. Commun. Rev. **5**(1), 3–55 (2001)
19. K. Sharp, F. Matschinsky, Translation of Ludwig Boltzmann's paper "on the relationship between the second fundamental theorem of the mechanical theory of heat and probability calculations regarding the conditions for thermal equilibrium" sitzungberichte der kaiserlichen akademie der wissenschaften. Mathematisch-naturwissen classe. abt. ii, lxxvi 1877,

pp 373–435 (wien. ber. 1877, 76: 373-435). Reprinted in wiss. Abhandlungen, vol. ii, reprint 42, p. 164–223, Barth, Leipzig, 1909. Entropy **17**(4), 1971–2009 (2015)

20. A.C. Sparavigna, Entropy in image analysis. Entropy **21**(5), 502 (2019)
21. J.L. Vazquez Noguera, H. Legal Ayala, C.E. Schaerer, J. Facon, A color morphological ordering method based on additive and subtractive spaces. In: *Image Processing (ICIP), 2014 IEEE International Conference on*. pp. 674–678. IEEE (2014)
22. M. Yousuf, M. Nobi, A new method to remove noise in magnetic resonance and ultrasound images. J. Sci. Res. **3**(1), 81 (2010)
23. M. Zavar, S. Rahati, M.R. Akbarzadeh-T, H. Ghasemifard, Evolutionary model selection in a wavelet-based support vector machine for automated seizure detection. Expert Syst. Appl. **38**(9), 10751–10758 (2011)

Deep Learning Models for Predicting COVID-19 Using Chest X-Ray Images

L. J. Muhammad [ID], Ebrahem A. Algehyne, Sani Sharif Usman [ID], I. A. Mohammed, Ahmad Abdulkadir, Muhammed Besiru Jibrin, and Yusuf Musa Malgwi

1 Introduction

Following the pandemicity of novel coronavirus disease 2019 (COVID-19 or 2019-nCoV), the first case of 2019-nCoV was reported on December 8, 2019, in Hubei Province, China, called Wuhan [1–3]. SARS-CoV-2 (severe acute respiratory syndrome coronavirus 2), the causative agent of COVID-19, has spread rapidly in the globe with the number of confirmed cases being 50,648,463 and 1,260,567 deaths were recorded as at November 8, 2020, 20:17 GMT [2–7]. On the basis of ribonucleotide sequence, SARS-CoV-2 belongs to the genus *Betacoronavirus* class [8, 9]. Seven different types of human coronaviruses (HCoVs) were recently reported to be members of *Alphacoronavirus* (229E and NL63 of HCoVs) and *Betacoronavirus*

L. J. Muhammad (✉) · M. B. Jibrin
Department of Mathematics and Computer Science, Faculty of Science, Federal University of Kashere, Gombe, Nigeria
e-mail: lawan.jibril@fukashere.edu.ng

E. A. Algehyne
Department of Mathematics, University of Tabuk, Tabuk, Saudi Arabia
e-mail: e.algehyne@ut.edu.sa

S. S. Usman
Department of Biological Sciences, Faculty of Science, Federal University of Kashere, Gombe, Nigeria
e-mail: ssu992@fukashere.edu.ng

I. A. Mohammed
Department of Computer Science, Yobe State University, Damaturu, Yobe, Nigeria

A. Abdulkadir
Kano University of Science and Technology, Wudil Kano, Nigeria

Y. M. Malgwi
Computer Science Department, Modibbo Adama University of Technology, Yola, Nigeria

© Springer Nature Switzerland AG 2022
P. Johri et al. (eds.), *Trends and Advancements of Image Processing and its Applications*, EAI/Springer Innovations in Communication and Computing, https://doi.org/10.1007/978-3-030-75945-2_6

(OC43, HKU1, SARS, MERS, and SARS-CoV-2 HCoVs) [3, 4, 8]. According to guidelines provided by the Chinese Diagnosis and Treatment Protocol for Novel Coronavirus Pneumonia, COVID-19 was investigated regardless of the results of reverse transcription polymerase chain reaction (RT-PCR) tests with chest computed tomography (CT) scan, pyrexia, cough, shortness of breath, lower respiratory tract infection, and lymphopenia among other laboratory results [3, 10, 11].

Although the mortality, spread, and transmission of COVID-19 potential are not fully understood, the epidemic of SARS-CoV-2 transmission significantly exceeds that of SARS-CoV [12, 13]. Therefore, the interaction of SARS-CoV-2 with ACE-2 and sialic-acid-containing glycoproteins and gangliosides is insufficient to detect potent upper respiratory tract infection [13–15]. The COVID-19 epidemic does not have a clinically proven antiviral drug; however, its patients recover with the help of antimicrobials, antimicrobials, and chloroquine and vitamin C supplementation [2–4, 14, 16]. As rapid testing of RT-PCR is becoming more and more appropriate, higher negative values, processing delays, and lower sensitivity, among others, remain [5, 14]. In addition, CT scans can be used to determine not only the various stages of COVID-19 detection and emergence but can also detect early COVID-19 in patients with poor RT-PCR testing or asymptomatic patients [17, 18]. CT scans were considered to be the first and most important tool in patients with suspected COVID-19 or confirmed cases at several facilities in Wuhan China and northern Italy, among others [17–20]. Chest CT scan in COVID-19 patients has found support especially in patients with traumatic breathing and in facilities with limited screening or in screening of patients with clinical manifestations and high probability of COVID-19 [17–21].

COVID-19 is the recent pandemic which affects the entire universe and its diagnostic technologies and treatments are still active under development and evaluation [22]. It is now evident that COVID-19 is now endemic and people must find a way to live with it till when its medications and vaccines are clinically tested, approved, and readily available in large quantity.

Owing to a dramatic increase in the number of both new and suspected COVID-19 cases, artificial intelligence (AI) approaches play a vital role for the prediction, detection, response, recovery of disease, and making clinical diagnosis or characterization of COVID-19 on imaging. The modern era largely depends on AI techniques such as machine learning (ML), data mining, and deep learning (DL), among others [23–26]. CT provides a clear-cut window for the aforementioned AI techniques as well as for the classification and quantification of COVID-19 disease [27, 28]. DL-driven models assist in the course of mitigating the pandemic like COVID-19 in terms of further spread, drugs and vaccine discovery, diagnosis, and patient care and treatment, among others [29, 30].

Application of DL which is one of the current flag-bearers of AI techniques nowadays for automatic diagnosis of the many diseases in the medical fields has recently gain popularity and becomes an adjuvant technique or a tool for clinicians and other medical health workers [28, 31–33]. DL techniques are used to develop end-to-end learning model in order to achieve the desired and optimum results using x-ray images without a need for manual extraction of the feature of the images. The techniques have been applied in the diagnosis, prediction, and detection of many diseases such as coronary artery disease detection, brain and breast cancer diagnosis, qne prediction of diabetes mellitus and even prediction, detection, and diagnosis of COVID-19. There is a need for experts of AI to harness DL techniques to develop simple, fast, and accurate DL-based diagnostic models and tools that might help clinicians and other health workers to mitigate COVID-19 pandemic or provide timely assistance to COVID-19 patients. Even though radiologists are playing very vital role in the course of mitigating the pandemic due to their vast experience and knowledge in the field, DL techniques assist them to obtain accurate data diagnosis [34]. The techniques also address the challenges such as limited readily available RT-PCR test kits, test cost, and waiting time of the test results. In this work, deep learning networks which include residual network 50, efficient network B4, and convolutional neural network are used and applied on x-ray chest dataset of patients suspected of having COVID-19 and those who appeared to be COVID-19 positive or have pneumonia including MERS, SARS, and ARDS to develop COVID-19 predictive models. The algorithms incorporate low-high-level image features and classifiers in an end-to-end multilayer manner for the level of the features to be magnified by the numbers of stacked layers.

2 Deep Learning

Machine learning (ML) is one of the artificial intelligence (AI) techniques that are used for the development of computational learning models, in which the models learn from experience using historic data [14, 29, 35]. But ML concept that is used to develop the models that recognize objects in images is called deep learning (DL). DL is also called deep structured learning or hierarchical learning, and it is one of the best inventions in the modern era of AI [36]. Until the 1990s, classical machine learning techniques were widely used in data processing, inference, and forecasting. Yet it had many drawbacks such as relying on handicrafts, tied to human-level accuracy and also sensitivity [25]. But in the case of DL, manual feature engineering is not required but rather features are extracted from the data during training. In addition, DL techniques make more accurate predictions and classification with the aid of innovative AI techniques, algorithms, computing power of modern machines, and available large dataset [37]. Learning presentation in conventional machine learning allows a machine to be fed with natural raw data and discover the learning representation needed for classification, detection, or prediction [38]. However, the learning presentation in DL allows many levels of presentation, obtained by composing nonlinear simple modules that each transform the representation at one level right from raw data input to a representation at a higher slightly more abstract level [38]. The theoretical foundation of DL is well rooted in the classical artificial neural network (NN), but unlike NN, DL uses many hidden neurons and its layers usually have two as architectural advantages combined with new training paradigms [39]. Resorting many neurons allows deep learning technique an extensive coverage of raw data at hand; layer-by-layer pipeline of nonlinear combination of their respective output generates a lower dimensional projection of the input space [39, 40]. Thus, every lower dimensional projection corresponds to a higher perpetual level, provided that the network is optimally weighted which resulted in an effective high-level abstraction of image or raw data. Deep learning allows the system that it built to be composed of many processing layers to learn representation of data with many levels of abstraction [39]. Therefore, DL is emerging nowadays as a powerful tool for machine learning, promising to reshape the future of AI. Deep learning can be defined as a machine learning approach that harnesses many layers of nonlinear data processing for either supervised or unsupervised feature extraction or transformation and for pattern classification and analysis [41, 42].

2.1 Convolutional Neural Network

Convolutional neural network (CNN) is the DL algorithm that has led to several breakthroughs for image classifications [38]. The structure of the CNN was proposed by Fukushima in 1988, but due to limited computational hardware for training the network, it was not widely used. However, when the gradient-based learning algorithm was applied to CNNs in the work of [43] for document recognition system, and the result was found to be successful, the researchers further improved the networks and reported the state-of-the-art results in many recognition systems [44, 45]. CNN is now dominant in various computer visions and other classification tasks and attracts interest across many application domains including radiology, and it became a method of choice in medical image analysis [46].

CNN is used for processing data that has a grid pattern, like pictures or images, inspired by the organization of the visual cortex of animals [47] and designed to automatically and flexibly learn the spatial distribution of features, from low to high patterns. CNN is composed of three types of layers or building blocks which include convolution, pooling, fully connected, and classification layers [46, 48]. Convolution and pooling layers are used for extraction of features, while fully connected layers are used for mapping the extracted features into final output layer which is usually called classification layer. High-level features are found in features that are propagated from low-level layers. As the features propagate to the very high level or layer, the size of the features decreases according to kernel size for convolutional and max-pooling operations, respectively [49, 50]. However, the number of feature maps usually increases by representing better features of input images to ensure the accuracy of the classification. The feed-forward neural networks are used as the classification layer as they have better performance [50]. In the classification layer, the extracted features are considered as inputs in relation to the dimension of weight of the final matrix neural network. Nowadays, global average pooling and global pooling techniques are used instead of fully connected layers, because they are more expensive in terms of learning and network parameters [50]. The convolution layer plays an important role in CNNs, as it is made up of a number of mathematical operations, such as convolution, a special type of linear functions or operations [46, 47].

Convolutional feature mapping from previous layer is convolved with learnable kernels; therefore, the output kernel go through nonlinear or linear activation functions such as hyperbolic tangent, sigmoid, and softmax, among others, to output feature maps [44]. As such, each feature map output is combined with more than one input feature map. As we can see in Eq. (1),

$$x_j^i = f\left(\sum_{i \in M_j} x_i^{l?1} ? k_{ij}^l + b_j^l \right),$$ (1)

where x_j^i is the output of the current layer, $x_i^{l?1}$ is the previous layer output, k_{ij}^l is the kernel for the present layer, and b_j^l are the biases for the current layer. M_j is the selection input map, while an addictive bias b is each output map.

At pooling layer the number of input and output features maps cannot change because if there are n input maps, then there must be n output map exactly. But the size of each dimension of the output map will be reduced due to the downsampling operation. The operation can be formulated in Eq. (2):

$$x_j^i = down\left(x_j^{l?1} \right).$$ (2)

The score of each extracted feature from convolutional layer in the preceding steps is computed at fully connected layer. The vector with scalar values is used to represent the final layer feature map and passed to fully connected layers. The fully connected feed-forward neural layers are used in the softmax classification layer.

The pixel values of digital images are stored in two-dimensional pair (2D) grid, i.e., a list or array of numbers, and a small parameter grid called a kernel, an optimizable extractor of feature, is applied to each area of the image, which makes CNNs work very well for image processing; because of its feature, it can be anywhere in the image. As one layer feeds its output into the next layer, the extracted elements can be hierarchically or sequentially and progressively be more complex [46, 49]. The process of optimizing barriers such as kernel is called training, which is done to reduce the difference between output and labels of the ground truth through an optimization algorithm called gradient descent and backpropagation [46, 48].

CNN architecture has an end-to-end structure, which is learn by high-level presentations from data [48]. CNN algorithms have also found great success in the application of image pathology such as the diagnosis of COVID-19, coronary artery diseases, diabetes, and cancers using high-resolution images.

In this work CNN is used and applied on x-ray chest dataset of patients who are suspected of having COVID-19 and those who are COVID-19 positive or have other viral and bacterial pneumonias (MERS, SARS, and ARDS) to develop CNN COVID-19 predictive model.

2.2   Residual Network

Residual network (ResNet) was developed by Kaiming in order to address the vanishing gradient problem that CNN predecessors had [51]. It had been developed with many different numbers of layers such as 34, 50, 101, 152, and even 1202 [44]. ResNet is a feed-forward CNN with residual connection. The residual layer is defined based on the output of $(I - 1)^{th}$ which drives from residual layer defines as x_{i-1}. $F(x_{i-1})$ is the output after performing various operations. The final output of the residual unit is x_I which can be defined in Eq. (3):

$$x_I = F\left(x_{i?1}\right) + x_{i?1}. \tag{3}$$

ResNet minimizes the vanishing gradient problem of CNN by allowing shortcut path for gradient to flow through, where identity mapping is used to bypass the CNN weight layer if the current is not necessary which helps avoid overfitting problem to the training set [30, 52]. In this work ResNet50 which contained 49 convolutional layers and 1 fully contented layer was used.

In addition ResNet50 is used and applied on x-ray chest dataset of patients who are suspected of having COVID-19 and those who are COVID-19 positive or have other viral and bacterial pneumonias (MERS, SARS, and ARDS) to develop ResNet50-based COVID-19 predictive model.

2.3   Efficient Networks

Efficient network (EfficientNet) was proposed by Tan and Le in 2019, and the network was based on the notion to balance all dimensions of network which include width, depth, and resolution [36]. EfficientNet is a lightweight CNN that achieved the state-of-the-art accuracy with an order of magnitude fewer parameters and FLOPs. Unlike other CNNs which scale the dimensions of network arbitrarily, efficient network which is a compounded scaling method uniformly scales network width, depth, and resolution with a set of fixed scaling coefficients.

The compounded scaling method of EfficientNet is very useful especially if the input image is bigger; the network needs more layers to increase the receptive field and channels to capture more fine-grained patterns on the bigger image [53]. EfficientNet uniformly scales network width, depth, and resolution with a set of fixed scaling coefficients in a principled manner using Eq. (4):

$$depth : d = ?^? \tag{4}$$

$$width : w = ?^? \tag{5}$$

$$resolution : r = ?^? \tag{6}$$

$$s.t. ?. ?^2 . ?^2 ? 2 \tag{7}$$

$$??1, ??1, ??1 \tag{8}$$

where α, β, γ are constants that can be determined by a small grid search, while $\emptyset$ is the user-defined coefficient that controls the quantity of the resource available for model scaling and α, β, γ define how the extra resources to network width, depth, and resolution are allocated [53].

EfficientNet relies on compound scaling to achieve greater performance without compromising the resource efficiency, and it is a mobile-size baseline network that uses different compound coefficients in the equation, ranging from EfficientNet B0 to EfficientNet B7 [54]. In this work EfficientNet is used and applied on x-ray chest dataset of patients who are suspected of having COVID-19 and those who are COVID-19 positive or have other viral and bacterial pneumonias (MERS, SARS, and ARDS) to develop EfficientNet B4 COVID-19 predictive model.

3 Related Work

Many works have employed DL networks for the development of predictive models for COVID-19 infections. As in [21], the predictive model for future cases of COVID-19 in Canada using the state-of-the-art DL LSTM algorithm was developed. The performance of the model was only evaluated using MSE error and accuracy and the model was found to have 34.83 RMSE errors and 93.4% accuracy. This study shows how DL models could help the Canadian government to monitor the situation of COVID-19 pandemic in Canada and also prevent further spread and transmission of the disease. LSTM, autoregressive integrated moving average (ARIMA), and prophet algorithm (PA) were used to predict the number of future confirmed, recoveries, and death cases of COVID-19 for the next 7 days in the work of [55]. The models showed a promising performance result with an average accuracy of 94.80% and 88.43% in Australia and Jordan, respectively. DL models can significantly help in the identification of the most infected cities in the two nations. Table 1 shows the recent works to mitigate COVID-19 pandemic. The DL techniques which include VGG-16, AlexNet, GoogleNet, ResNet18, ResNet50, ResNet101, SqueezeNet, MobileNet, and Xception were used to manage COVID-19 in routine clinical practice using computed tomography image dataset in the work of [56]. Among the all techniques, ResNet101 and Xception achieved the best performance. ResNet101 has achieved 0.994 AUC score, 99.02% specificity, 100% sensitivity, and 99.51% accuracy, while Xception has achieved 0.994 AUC score, 98.04% specificity, 98.04% sensitivity, and 99.02% accuracy. COVID-19 computer-aided detection from x-ray images was developed with multi-CNN and Bayes net classifier in the work of [57]. The approach adopted in the study combines features extracted from multi-CNN with correlation-based feature selection techniques and Bayes net classifier for COVID-19 detection. The experimental result of the study showed the efficiency of pre-trained multi-CNN over a single CNN for COVID-19 detection with 0.963 AUC score and 91.16% accuracy. DL

Table 1 Dataset description

SN	Attribute	Value	Description
1	Index	Interger	Primary identifier
2	X_ray_image_name	Unique values (IM-0128-0001.Jpeg)	Image name of the x-ray, few image names contain virus or bacteria tag
3	Image_Labels	Pneumonia Normal	It indicates x-ray is normal or healthy and the person affected with pneumonia
4	Data_type	TRAIN TEST	It describes where x-ray image belongs to train or test set
5	Label_2_Virus_category	[null] COVID-19 Others	The label holds the information about whether pneumonia is due to virus, bacteria, or ARDS
6	Label_1_Virus_categorysort	Bacteria [null] Others	The label holds the information about pneumonia (virus, bacteria, or ARDS)

networks have been used to characterize COVID-19 pneumonia in chest images in the work of [58]. The DL learning network helps the radiologists to make fast and quicker diagnosis of COVID-19. In the work of [59], the model for prediction of the severity of COVID-19 pneumonia with DL network using chest x-ray was developed. It was able to gauge the severity of COVID-19 lung infection for care and treatment especially to patients that are in the ICU unit of hospitals. In the work of [3], supervised machine learning models using support vector machine, decision tree, logistic regression, neutral network, and naïve Bayes algorithms with COVID-19 epidemiology labeled dataset for positive and negative cases in Mexico.

4 Methods and Materials

This section provides dataset description and processing, analysis method, and CNN, ResNet50, and EfficientNet B4 algorithms for development of COVID-19 predictive models. Figure 1 shows the methodology adopted for the work.

4.1 Dataset

X-ray chest dataset of patients who are suspected of having COVID-19 and those who are COVID-19 positive or have other viral and bacterial pneumonias (MERS, SARS, and ARDS) was sourced from public sources as well as through indirect collection from hospitals and physician and released publicly in the GitHub repository (https://github.com/ieee8023/covid-chestxray-dataset) [60]. The project of the collection of the dataset was approved by the University of Montreal's Ethics Committee #CERSES-20-058-D.

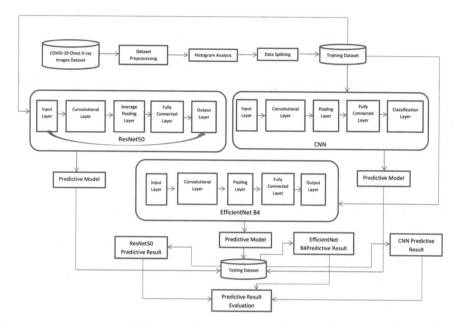

Fig. 1 General workflow of the methodology

	Unnamed: 0	X_ray_image_name	Label	Dataset_type	Label_2_Virus_category	Label_1_Virus_ca
0	0	IM-0128-0001.jpeg	Normal	TRAIN	NaN	NaN
1	1	IM-0127-0001.jpeg	Normal	TRAIN	NaN	NaN
2	2	IM-0125-0001.jpeg	Normal	TRAIN	NaN	NaN
3	3	IM-0122-0001.jpeg	Normal	TRAIN	NaN	NaN
4	4	IM-0119-0001.jpeg	Normal	TRAIN	NaN	NaN

Fig. 2 Sample of the dataset

4.2 Data Preparation

The dataset has 5910 instances with 6 attributes which include index, X_ray_image_
name, Image_Labels, Data_type, Label_2_Virus_category, and Label_1_Virus_cat-
egorysort. Table 1 shows the value and description of the dataset attributes (Fig. 2).

In the dataset all the COVID-19 patients were classified as having pneumonia;
thus, none them is classified as normal. More so, in target "Label_2_Virus_cate-
gory," the "unknown" value is associated with the majority of images as such
unknown values consist of 98.7% of the total cases, while COVID-19 values consist
of less than 1.3% of the total cases. Therefore, training a model to classify

Label_2_virus_category with 98.7% accuracy, it will be highly inefficient in detecting true positive COVID-19 cases. We only consider instances of the dataset with the label normal and pneumonia + COVID-19 cases so as to develop models which differentiate between COVID-19 patients and normal patients. However, some of the chest x-ray images are not squared but automatically reside the images to 356 * 256 pixels using the techniques proposed by [61] in order to get and maintain the much-needed information of the region of interest in chest x-ray images.

4.3 Data Augmentation

DL networks heavily rely on high-quality training sample in order to achieve or get accurate prediction, detection, and classification results from the dataset [22]. Therefore, DL networks need a large amount of training data build good performing models, the training set is small. Therefore, in this work image augmentation random rotational technique was used in order to boost the performance of the DL networks. The images were augmented with random rotational technique by 90 degrees range, width_shift_range = 0.15, height_shift_range = 0.15, horizontal_flip = True, zoom_range = [0.9, 1.25], and brightness_range = [0.5, 1.5]. All the networks which include ResNet50 and EfficientNet B4 were fed with the resized chest x-ray images of 224 * 224 pixels which were done automatically, in order to get and maintain the much-needed information of the region of interest of the images. The sigmoid function (f) was used for networks as the activation function as shown in Eq. 3.

$$f(x) = Sigmoid(x)\frac{1}{1+e^{?x}} \qquad (9)$$

Random numbers were set for the number of neurons, weights, and bias for the classification process for all networks which include CNN, ResNet50, and EfficientNet B4.

4.4 Environment Setup

Python which is an open-source programming was used for the development of the models using Corei7, GTX1060 6GB GPU computer system. Python Notebook was used as the development environment and all the necessary libraries were installed such as keras, tensor, numpy, pandas, matplotlib, seaborn, tensorflow, and efficientnet, among others.

5   Results and Discussion

COVID-19 predictive models using chest x-ray images were developed with CNN, ResNet50, and EfficientNet B4 deep learning algorithms. Out of 6559 instances of the dataset, only consider 852 instances of the dataset with the label normal and pneumonia + COVID-19 cases so as to develop models which differentiate between COVID-19 patients and normal patients. The dataset was trained on each algorithm with 80% training data and the remaining 20% as a testing dataset. A sample of the chest x-ray images of training and testing datasets is shown in Figs. 3 and 4, respectively.

Likewise, the chest x-ray images having classified as pneumonia + COVID-19 (COVID-19 patients) while the chest x-ray images having classified as normal (healthy patients) histograms are shown in Figs. 5 and 6, respectively.

6   Performance Evaluation

The COVID-19 predictive models such as CNN, ResNet50, and EfficientNet B4 were evaluated based on the following performance evaluation metrics for deep learning algorithms:

(i) Accuracy is the learning evaluation metric that is used to determine or measure the percentage of the instances of the x-ray images of the dataset that were correctly predicted with the learning network model. Accuracy is defined using the below equation:

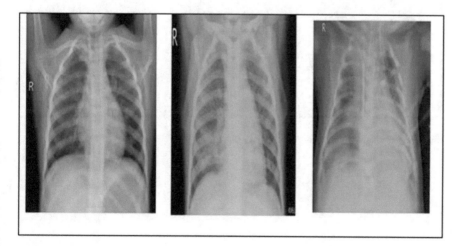

Fig. 3 Sample of chest x-ray image training dataset

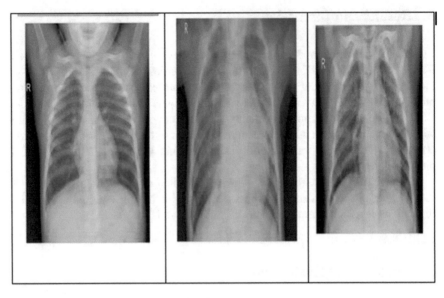

Fig. 4 Sample of chest x-ray image testing dataset

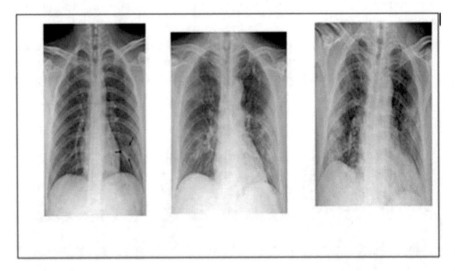

Fig. 5 Sample of x-ray images of COVID-19 patients

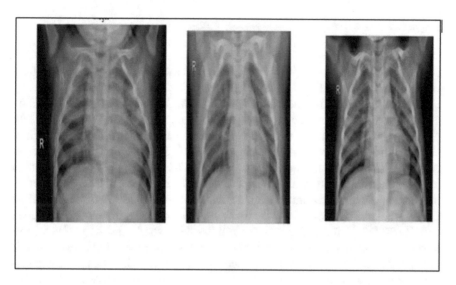

Fig. 6 Sample of x-ray images of non-COVID-19 patients

$$Accuracy = \frac{tp + tn}{tp + tn + fn + fp} \tag{10}$$

(ii) Sensitivity is the learning evaluation metric that is used to determine or measure the percentage of COVID-19 positive cases from the x-ray images of the dataset that were correctly predicted with the learning network model. The sensitivity is defined using the below equation:

$$Sensitivity = \frac{tp}{tp + fn} \tag{11}$$

(iii) Specificity is the learning evaluation metric that is used to determine or measure the percentage of COVID-19 negative cases from the x-ray images of the dataset that were correctly predicted with the learning network model. The specificity is defined using the below equation:

$$Specificity = \frac{tp}{tn + fp} \tag{12}$$

(iv) The area under the curve (AUC) is the learning evaluation metric that is used to determine the percentage or measure the score of how much the model differentiates between COVID-19 positive cases and COVID-19 negative cases. The AUC is defined using the below equation:

$$AUC = \frac{fp}{tn + fp} \tag{13}$$

Note: *tp* stands for true positive, *tn* stands for true negative, *fp* stands for false positive, and *fn* stands for false negative.

Each of the predictive models developed in this study was evaluated based on accuracy, sensitivity, and specificity, and AUC evaluations metrics. The COVID-19 predictive model developed with ResNet50 deep learning network has the highest accuracy of 99% for predicting instances of COVID-19 cases from the x-ray images correctly, followed by the model developed with CNN deep learning network which has a 96% accuracy and then the model developed with EfficientNet B4 deep learning network which has the lowest accuracy of 94%. Likewise for predicting COVID-19 positive cases from the x-ray images correctly, the model developed with ResNet50 deep learning network still has the highest sensitivity of 89%, followed by the model developed with EfficientNet B4 deep learning network which has an 86% sensitivity and then the model developed with CNN deep learning network which has the lowest sensitivity of 78%. For predicting COVID-19 negative cases from the x-ray images correctly, the model developed with EfficientNet B4 deep learning network has the highest specificity of 89%, followed by the model developed with CNN deep learning network which has an 86% specificity and then the model developed with ResNet50 deep learning network which has the lowest specificity of 69%. Moreover, in terms of how much the model differentiates between COVID-19 positive cases and COVID-19 negative cases, the model developed with ResNet50 deep learning network still has the highest score of 96% AUC, followed by the model developed with CNN deep learning network which has a 92% AUC score and then the model developed with EfficientNet B4 deep learning network which has the lowest AUC score of 73.50. Table 2 shows the performance evaluation results of the models and Fig. 7 shows the chart presentation of the performance evaluation results of the models. Therefore, ResNet50 model with the highest sensitivity of 89.50% is used to diagnose COVID-19 infection and is used as an adjuvant tool in radiology department in hospitals.

Table 2 Performance evaluation results of the models

S/N	Model	Accuracy (%)	Sensitivity (%)	Specificity (%)	AUC (%)
1	CNN	96.00	78.50	84.00	92.00
2	ResNet50	99.00	89.50	69.00	96.50
3	EfficientNet B4	94.50	86.00	89.00	73.50

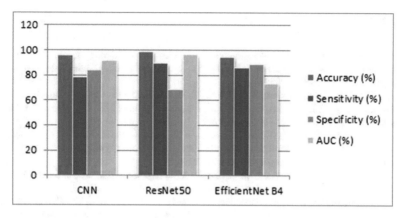

Fig. 7 Chart presentation of the performance evaluation results of the models

7 Conclusion

Since the inception of the first wave of COVID-19 pandemic, lots of lives have been lost and many businesses were shut down which made most of the countries went into economic recession. The vaccines against the pandemic are now being developed with uncommon commitments and substantial efforts are being made by various scientists and researchers around the world. However, immediately after the first wave of COVID-19 pandemic, the second wave of the pandemic has now started and causes a lot of lives and grounds a lot of businesses that have resumed around the world and yet most of the countries are still facing inadequate and standard laboratories for testing a large number of suspected COVID-19 cases. In this study, deep learning algorithms were used to develop models for predicting COVID-19 using chest x-ray images, and the models were able to extract COVID-19 imagery features and provide clinical diagnosis ahead of the pathogenic test in order to save time and complement COVID-19 testing laboratories. The ResNet50-based model developed in this study was found to have the highest accuracy of 99%, followed by the CNN-based model which has a 96% accuracy and then the EfficientNet B4-based model which has the lowest accuracy of 94%. The ResNet50-based model still has the highest sensitivity of 89%, followed by the EfficientNet B4-based model which has an 86% sensitivity and then the CNN-based model which has the lowest sensitivity of 78%. EfficientNet B4 was found to have the highest specificity of 89%, followed by the CNN-based model which has an 86% specificity and then ResNet50-based model which has the lowest specificity of 69%. ResNet50 deep learning network still has the highest score of 96% AUC, followed by the CNN-based model which has a 92% AUC score and then EfficientNet B4-based model which has the lowest AUC score of 73.50. Therefore, ResNet50 model with the highest sensitivity of 89% is used for the diagnosis of COVID-19 infection as well as an adjuvant tool in radiology department in hospitals.

References

1. M. Islam, S. Mahmud, L.J. Muhammad, et al., Wearable technology to assist the patients infected with novel coronavirus (COVID-19). SN Comput. Sci. **1**, 320 (2020). https://doi.org/10.1007/s42979-020-00335-4
2. L.J. Muhammad, M.M. Islam, S.S. Usman, et al., Predictive data mining models for novel coronavirus (COVID-19) infected patients' recovery. Springer Nat. Comput. Sci. 1 (2020). https://doi.org/10.1007/s42979-020-00216-w
3. L.J. Muhammad, E.A. Algehyne, S.S. Usman, et al., Supervised machine learning models for prediction of COVID-19 infection using epidemiology dataset. Springer Nat. Comput. Sci. (2020). https://doi.org/10.1007/s42979-020-00394-7
4. L.J. Muhammad, S.S. Usman, Power of artificial intelligence to diagnose and prevent further COVID-19 outbreak: a short communication. arXiv:2004.12463 [cs.CY] (2020)
5. S.M. Ayyoubzadeh, S.M. Ayyoubzadeh, H. Zahedi, et al., Predicting COVID-19 incidence through analysis of Google trends data in Iran: data mining and deep learning pilot study. JMIR Public Health Surveill. **6**(2), e18828 (2020)
6. C. Huang, et al., Clinical features of patients infected with 2019 novel coronavirus in Wuhan, China. Lancet 56 - 63, (2020) ahead of print
7. https://www.worldometers.info/coronavirus/. Accessed 8th November, 2020
8. D.S. Hui, E.I. Azhar, T.A. Madani, et al., The continuing 2019-nCoV epidemic threat of novel coronaviruses to global health – the latest 2019 novel coronavirus outbreak in Wuhan. China. Int. J. Infect. Dis. **91**, 264–266 (2020)
9. I.M. Ibrahim, D.H. Abdelmalek, M.E. Elshahat, et al., COVID-19 spike-host cell receptor GRP78 binding site prediction. J. Infect. **80**(5), 554–562 (2020)
10. Chinese diagnosis and treatment plan of COVID-19 patients (The fifth edition). http://www.nhc.gov.cn/yzygj/s7653p/202002/3b09b894ac9b4204a79db5b8912d4440.shtml. (2020a)
11. Chinese diagnosis and treatment plan of COVID-19 patients (The sixth edition). http://www.nhc.gov.cn/yzygj/s7653p/202002/8334a8326dd94d329df351d7da8aefc2.shtml. (2020)
12. X. He et al., Temporal dynamics in viral shedding and transmissibility of COVID-19. Nat. Med. (2020). https://doi.org/10.1038/s41591-020-0869-5
13. W. Li, M.J. Moore, N. Vasilieva, et al., Angiotensin–converting enzyme 2 is a functional receptor for the SARS coronavirus. Nature **426**, 450–454 (2003)
14. D. Wang, B. Hu, C. Hu, et al., Clinical characteristics of 138 hospitalized patients with 2019 novel coronavirus-infected pneumonia in Wuhan, China. JAMA (2020). https://doi.org/10.1001/jama.2020.1585
15. V.K.R. Chimmula, L. Zhang, Time series forecasting of COVID-19 transmission in Canada using LSTM networks. Chaos, Solitons Fractals **109864 -109873** (2020)
16. L. Dong, S. Hu, J. Gao, Discovering drugs to treat coronavirus disease 2019 (COVID-19). Drug Discov. Ther. **14**, 58–60 (2020)
17. S. Jin, et al., AI-assisted CT imaging analysis for COVID-19 screening: building and deploying a medical AI system in four weeks. medRxiv http://www.medrxiv.org/content/10.1101/2020.1103.1119.20039354v20039351 (2020)
18. Y. Li, L. Xia, Coronavirus disease 2019 (COVID-19): role of chest CT in diagnosis and management. Am. J. Roentgenol. **1–7** (2020). https://doi.org/10.2214/AJR.20.22954
19. W. Zhao, Z. Zhong, X. Xie, et al., Relation between chest CT findings and clinical conditions of coronavirus disease (COVID-19) pneumonia: a multicenter study. Am. J. Roentgenol. **1–6** (2020). https://doi.org/10.2214/AJR.20.22976
20. N. Sverzellati et al., Integrated radiologic algorithm for COVID-19 pandemic. J. Thorac. Imaging (2020). https://doi.org/10.1097/RTI.0000000000000516
21. G.D. Rubin et al., The role of chest imaging in patient management during the COVID-19 pandemic: a multinational consensus statement from the Fleischner society. Chest (2020). https://doi.org/10.1148/radiol.2020201365

22. R. Zhang, Z. Guo, Y. Sun, et al., COVID19XrayNet: a two-step transfer learning model for the COVID-19 detecting problem based on a limited number of chest X-ray images. Interdiscip Sci Comput Life Sci. (2020). https://doi.org/10.1007/s12539-020-00393-5
23. H. Sadiq, L.J. Muhammad, A. Yakubu, Mining social media and DBpedia data using Gephi and R. J. App. Comput. Sci. Math. **12**(1), 14–20 (2018)
24. A.A. Haruna, L.J. Muhammad, B.Z. Yahaya, et al. An improved C4.5 data mining driven algorithm for the diagnosis of coronary artery disease. In *International Conference on Digitization (ICD)*, Sharjah, United Arab Emirates, p 48–52 (2019)
25. S. Hussain et al., Performance evaluation of various data mining algorithms on road traffic accident dataset, in *Information and Communication Technology for Intelligent Systems*, Smart Innovation, Systems and Technologies, ed. by S. Satapathy, A. Joshi, Singapore: Springer; (2019), p. 106
26. L.J. Muhammad, E.A. Algehyne, S.S. Usman, Predictive supervised machine learning models for diabetes mellitus. SN Comput. Sci. **1**, 240 (2020). https://doi.org/10.1007/s42979-020-00250-8
27. F.S. Ishaq, L.J. Muhammad, B.Z. Yahaya, et al., Fuzzy based expert system for diagnosis of diabetes mellitus. Int. J. Adv. Sci. Technol. **136**, 39–50 (2020)
28. L.J. Muhammad, E.J. Garba, N.D. Oye, et al., On the problems of knowledge acquisition and representation of expert system for diagnosis of coronary artery disease (CAD). Int. J. u-e-Serv. Sci. Technol. **11**(3), 50–59 (2018)
29. H. Kaiming, Z. Xiangyu, R. Shaoqing, et al., Deep residual learning for image recognition. arXiv:1512.03385 [cs.CV] (2015)
30. F. Shi, J. Wang, J. Shi, et al., Review of artificial intelligence techniques in imaging data acquisition, segmentation and diagnosis for covid-19, arXiv preprint arXiv:2004.02731 (2020)
31. F.S. Ishaq, L.J. Muhammad, B.Z. Yahaya, et al., Data mining driven models for diagnosis of diabetes mellitus: a survey. Indian J. Sci. Technol. **11**, 42 (2018)
32. L.J. Muhammad, et al., Performance evaluation of classification data mining algorithms on coronary artery disease dataset. In *IEEE 9th International Conference on Computer and Knowledge Engineering (ICCKE 2019)*, Ferdowsi University of Mashhad (2019)
33. L.J. Muhammad et al., Using decision tree data mining algorithm to predict causes of road traffic accidents, its prone locations and time along Kano –Wudil Highway. Int. J. Database Theor. Appl. **10**, 11197–11208 (2017)
34. O. Tulin, T. Muhammed, A.Y. Eylul, et al., Automated detection of COVID-19 cases using deep neural networks with X-ray images. Comput. Biol. Med. **121**, 103792 (2020)
35. L.J. Muhammad, E.J. Garba, N.D. Oye, G.M. Wajiga, A.B. Garko, Mining framework to knowledge acquisition for expert system – a study on coronary artery disease, in *Translational Bioinformatics in Healthcare and Medicine*, Advances in Ubiquitous Sensing Applications for Healthcare, vol. 13, (Academic Press, Elsavier 2021), p. 1
36. T. Mingxing, V.L. Quoc, EfficientNet: rethinking model scaling for convolutional. Neural Networks, arXiv **1905**, 11946 (2019)
37. C. Tan, F. Sun, T. Kong, A survey on deep transfer learning, in *International Conference on Artificial Neural Networks*, (Springer, 2018), pp. 270–279
38. Y. LeCun, L. Bottou, G.B. Orr, et al., Efficient backprop, in *Neural Networks: Tricks of the Trade*, (Springer India, 1998), pp. 9–50
39. D. Ravì et al., Deep learning for health informatics. IEEE J. Biomed. Health Inform. **21**(1), 4–21 (2017)
40. L.J. Muhammad, B.Z. Yahaya, A. Garba, et al., Multi query optimization algorithm using semantic and heuristic approaches. Int. J. Database Theor. Appl. **9**(6), 219–226 (2016)
41. L. Deng, D. Yu, Deep learning: methods and applications. Found. Trends Signal Process. **7**(3–4), 197–387 (2013)
42. H. Wu, S. Prasad, Semi-supervised deep learning using Pseudo labels for hyperspectral image classification. IEEE Trans. Image Process. **27**(3), 1259–1270 (2018)

43. Y. LeCun, L. Bottou, Y. Bengio, et al., Gradient-based learning applied to document recognition. Proc. IEEE **86**, 2278–2324 (1998)
44. M.Z. Alom, T.M. Taha, C. Yakopcic, et al., A state-of-the-art survey on deep learning theory and architectures. Electronics **8**, 292 (2019)
45. L.J. Muhammad, A. Garba, G. Abba, Security challenges for building knowledge based economy in Nigeria. Int. J. Secur. Appl. **9**(1), 119 (2015)
46. R. Yamashita, M. Nishio, R.K.G. Do, et al., Convolutional neural networks: an overview and application in radiology. Insights Imaging **9**, 611–629 (2018). https://doi.org/10.1007/s13244-018-0639-9
47. D.H. Hubel, T.N. Wiesel, Receptive fields and functional architecture of monkey striate cortex. J. Physiol. **195**, 215–243 (1968)
48. M.C. Chen, R.L. Ball, L. Yang, et al., Deep learning to classify radiology free-text reports. Radiology **286**, p845–p852 (2018)
49. V. Nair, G.E. Hinton, Rectified linear units improve restricted Boltzmann machines. In *Proceedings of the 27th International Conference on Machine Learning (ICML-10)*, Haifa, Israel, 21–24 June 2010, pp. 807–814 (2010)
50. G.E. Hinton, S. Osindero, Y.-W. Teh, A fast learning algorithm for deep belief nets. Neural Comput. **18**, 1527–1554 (2006)
51. H. Choi, S. Ryu, H. Kim, Short-term load forecasting based on ResNet and LSTM. In *2018 IEEE International Conference on Communications, Control, and Computing Technologies for Smart Grids (SmartGridComm)*, Aalborg, pp. 1–6, https://doi.org/10.1109/SmartGridComm.2018.8587554 (2018)
52. D. Theckedath, R.R. Sedamkar, Detecting affect states using VGG16, ResNet50 and SE-ResNet50 networks. SN Comput. Sci. (2020). https://doi.org/10.1007/s42979-020-0114-9
53. M. Tan, B. Chen, R. Pang, et al, MnasNet: platform-aware neural architecture search for mobile. CVPR (2019)
54. T. Jebara, *Machine Learning: Discriminative and Generative* (Springer, Norwell, 2003)
55. A. Alazab, A. Awajan, A. Mesleh, et al., COVID-19 prediction and detection using deep learning. Int. J. Comput. Inform. Syst. Industr. Manag. Appl. **12**(2020), 168–181 (2020) ISSN 2150–7988
56. A.A. Ali, R.K. Alireza, A.U. Rajendra, et al., Application of deep learning technique to manage COVID-19 in routine clinical practice using images: results of 10 convolutional neural networks. Comput. Biol. Med. 1–9, (2020)
57. B. Abraham, M.S. Nair, Computer-aided detection of COVID-19 from X-ray images using multi-CNN and Bayesnet classifier. Biocybern. Biomed. Eng. (2020). https://doi.org/10.1016/j.bbe.2020.08.005
58. N. Qianqian, Y.S. Zhi, Q. Li, et al., A deep learning approach to characterize 2019 coronavirus disease (COVID-19) pneumonia in chest CT images. Eur. Radiol. (2020). https://doi.org/10.1007/s00330-020-07044-9
59. J. Cohen, L. Dao, K. Roth, et al., Predicting COVID-19 pneumonia severity on chest X-ray with deep learning. Cureus **12**(7), e9448 (2020). https://doi.org/10.7759/cureus.9448
60. https://github.com/ieee8023/covid-chestxray-dataset. Accessed 28th August, 2020
61. F. Pasa, V. Golkov, F. Pfeier, et al., Efficient deep network architectures for fast chest X-ray tuberculosis screening and visualization. Sci. Rep. **9**, 6268 (2019)

Deep Learning Methods for Chronic Myeloid Leukaemia Diagnosis

Tanya Arora, Mandeep Kaur, and Parma Nand

1 Scope of Study

The scope of this chapter explains the extent to which the research area of the study is explored and the parameters to be studied are specified in the study. Deep learning is an emerging and exceptional subfield that can manage large quantities of data efficiently and produce more accurate results.The study in this chapter compares the approaches for machine learning algorithms with the advanced methods of deep learning architecture taking into consideration the variants of artificial neural networks, further highlighting the concept of convolutional neural network (CNN) for detection, extraction, segmentation, and classification of the steps of diagnosis of chronic myeloid leukaemia like pre-processing, image segmentation, post-processing, and pattern analysis among others. An architecture has been designed for deep learning which uses the concept of CNN, being the centre of attraction and the most computational model among the other variants of ANN. The utilization of deep learning has been conducted for the implementation of the algorithm so as to improve the accuracy and the efficacy for better performances. The study population includes scientists, experts, students with MSc and PhD, postdocs, and everyone interested in the topics covered. This chapter could be used as a reference guide for artificial intelligence, medical, and biomedical education courses.

T. Arora · M. Kaur (✉) · P. Nand
Department of Computer Science & Engineering, School of Engineering and Technology (SET), Sharda University, Greater Noida, UP, India
e-mail: mandeep.kaur@sharda.ac.in; Parma.nand@sharda.ac.in

© Springer Nature Switzerland AG 2022
P. Johri et al. (eds.), *Trends and Advancements of Image Processing and its Applications*, EAI/Springer Innovations in Communication and Computing, https://doi.org/10.1007/978-3-030-75945-2_7

1.1 Literature Survey

At the National Level

The authors distinguished [1] different biosensors that are designed for diagnosing nucleic acid based on protein biomarkers for cancers.

The biomarker detection in cancer uses different approaches like electrochemical, optical, and mass-based transduction systems.

These features detailed performances of different biosensor designs analytically related to cancer biomarkers.

Future Challenges:
- Better sensitivity
- Analysis in real time
- Cost-efficient
- Mutiplex detection
- Integration of microfluidic technologies

The author described [2] the advancement in microfluidics and included lab-on-a-chip fabrication technique designs to minimize the standard cytometer by making it more cost-efficient.

This device does not require hard-core professionals to operate the device and can be operated by itself without technical manpower professionals.

A single-usable chip which is credit card sized is placed in this device having reagents, where the sample is filled and is insulated. The cartridge controller, which is as big as the size of a toaster, performs the handling of fluid and resistive measurement to provide the results in minutes.

Future Challenges:
- Use of nano-technology could yield more accurate and appropriate results in accordance with the ease of handling of point of care device

At the International Level

The author [3] evaluated a statistical predictive model for something like the categorization of survival of CML patients using algorithms for decision trees, in order to get access to variables diagnostic for survival of CML. The Obafemi Awolowo University Teaching Hospital Complex (OAUTHC), Ile-Ife, Southwest Nigeria, gathered historical range of set of data supposed to contain information about the variables observed during imatinib care follow-up. The predictive modelling approach for CML was developed using algorithms for the decision trees C4.5 and CART and simulated using the WEKA framework. The efficiency of the predictive model was assessed using the gathered sets of data through ten-fold cross-validation.

Future Challenges:
- Collection of more data about CML patients who did not survive CML in 2 years.
- The predictive developed model for 2-year CML survival can be incorporated into various Health Information Systems (HIS) for monitoring and real-time analysis of the medical data of CML patients which could also enable experts to start providing treatment strategies depending on the result of the CML patient survival categorization.

The authors discussed [4] functionalized nano-sensors and bio-chips for early and hypersensitive diagnosis of novel biomarkers for cancer.

Future Challenges:
- Genetic heterogeneity.
- For serum biomarker detection, high sensitivity and specific non-invasive tool is required.

The authors presented [5] the scientific advances on detecting bio-sensors of DNA for BCR-ABL fusion gene. The assembling of nano-structured programme is discussed with molecular action plan that is immovable and the performances of the devices are also analysed that use electrochemical and even optical methods for the detection of BCR-ABL fusion gene and the recyclability of geno-sensors for BCR-ABL fusion gene. The nano-structured platform is considered as a favourable instrument for the detection and keeps track of the BCR-ABL fusion gene in patients with leukaemia.

Future Challenges:
- Development of Smart Biosensor could facilitate early diagnosis of cancer.
- Refinement in the quality of life in leukaemic patients.

The authors revealed [6] the biological biomarkers like ctDNA, CTCs, and exosomes, methods for their enrichment and detection, and their potential for clinical applications.

Future Challenges:
- Interpretation of data which is derived from multi-marker investigations.
- Smartphone-based strategies could further monitor cancer at the point of care towards personalized medicine.
- Small cohorts of patients and large-scale investigations required to confirm liquid biopsy capabilities.
- Emerging lab-on-a-chip technologies towards integrated smart methodologies may accelerate liquid biopsy outcomes.

The authors developed [7] a multi-functional chip which was tested utilizing gold nano-probes so that RNA optical detection is performed within microfluidic chip that deploys the requirement of molecular amplification steps. For instant detection of CML, this device was used. The target RNAs from samples were mixed with gold nano-probes and saline solution in the chip to conclude and deliver a result with the help of the final colorimetric properties.

Precise output results were shown within 3 minutes in the trials with SNR up to 9 dB. In comparison with the results of the previous research, these results at micro scale were at least ten times quicker as compared to the reported time for CML methods for detection.

Future Challenges:
- Integrating microfluidic programme into a stand-alone device.
- Automatic and more cost-efficient generic DNA/RNA tests could be appropriate for POC screening.
- Easy follow-up with patients.

The author [8] provided a detailed overview of the widely applied deep learning algorithms and explored its implementations for the 'smart' output. Initially they discussed the evolution of deep learning techniques and about their potential benefits paired with the supervised machine learning. Consequently, computational researchers concentrated upon deep learning introduced explicitly the goal of enhancing the development of device efficiency. Some prominent models of deep learning are addressed in a very significant manner. Eventually, new research topics related to deep learning are illustrated, and some future developments along with concerns for smart manufacturing relevant to deep learning have been discussed.

Future Challenges:
- While computing capabilities like cloud computing, fog computing, etc. develop, computational intelligence techniques, like deep learning, can be moved across into the cloud, allowing more efficient as well as on-demand computing capabilities for smart development.

The authors described [9] an in vivo aphaeretic CTC isolation system that is indwelling intravascularly, which continuously collects circulating tumour cells, straightaway at the point from a peripheral vein. The leftover blood products are returned by the system after these cells are collected.

Future Challenges:
- Could be non-invasive.
- Circulating tumour cells could be dispersed at a continuous rate.
- More reliable.
- Higher recovery rate.

The authors presented [10] a practical and reliable tool for identification of biological materials, called electrochemical impedance spectroscopy based on simulation such as cancer biomarkers. Various variants of biomarkers and bio-sensor designs are used to detect cancer. Out of which, many are able to diagnose biomarkers in a broad range from $fg \, mL-1- \, ng \, mL-1$.

Future Challenges:
- Microfluidics could make such devices of great potential.
- Use of nanomaterial could be more accurate and practical in the developing sensors for biomarker detection.

2 Introduction

Cancer is the world's third leading principal cause of death. Both researchers and physicians face cancer-fighting challenges. Early cancer detection is a primary concern in saving many people's lives. For cancer diagnosis, visual observation and manual approaches are usually used. This manual processing medicinal images requires significant time and is particularly susceptible to errors.

To this effect, computer-aided diagnostic (CAD) systems have been proposed in the early 1980s to help support healthcare professionals to increase the quality of analysis of medical images [11]. Deep learning is the machine learning approach used for implementation.

Deep learning is an emerging field of machine learning which now has become increasingly common in today's world. Deep learning is the term often used to refer to the architectures that include several hidden layers called the deep networks for learning various features with several abstraction levels. Conventional and traditional machine learning methods are limited to the natural data which is further processed in its previous raw form. Installation of a machine learning system has been a major need for decades. Deep learning enables the input of raw data in the form of image pixels to the training algorithm before even retrieving the features or specifying a vector function. Algorithms of deep learning and methods can follow the correct set of features, and it does so in a much viable way than just using hand-coding to retrieve these functions. Rather than crafting with hand a set of conventional standards and the algorithms to retrieve features from the obtained raw data, deep learning criteria include automatically learning these simple features even during the stage of training. In deep learning, a challenge is recognized in the context of hierarchical concept, with each one being constructed on top of one another. The values of pixel intensity are given as some inputs into deep learning with a picture setup. A variety of hidden layers then remove the functionality from the image data. In a hierarchical fashion, these secret layers are built upon one another. Initially the network's lower-level layers trace only the regions which are edge based. Then, these same regions with edges are being used for identifying corners (intersection of edges) and contours (object outlines). The higher-level layers incorporate corners and contours to proceed in the next layer to more specific 'object sections'. The major characteristic of deep learning is that these featuring layers aren't really customized and developed by human engineers; on the contrary, they are slowly learnt from data through a general learning process. Lastly, the output layer categorizes the image and gets the output training sample—the data which is produced at the output layer seems to be deeply affected by other nodes in the network available so far. This method can be regarded as training with hierarchy because each level in this network takes the output of the preceding layer, making 'building blocks' to create ever more complicated concepts at the higher levels. Figure 1 contrasts conventional machine learning approaches with deep learning approaches which are further based on hierarchical representation of learning based on hand-crafted features (Fig. 2).

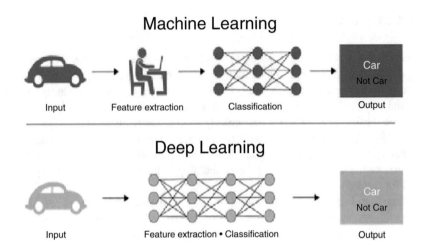

Fig. 1 Machine learning vs deep learning [11]

2.1 Why Use Deep Learning

Machine learning experts have invested an incredible time collecting knowledge software features. The machine learning algorithms have already taken decades, of so-called human efforts, at the time the big bang introductional approach of deep learning, which accumulates appropriate range of features that are considered necessary to categorize the input. Deep learning had already overcome those machine learning algorithms in terms of accuracy as the functions are learnt from the data using only a learning method with a general purpose rather than what human engineers design. Deep networks proved spectacular advancements in computing services that have already enhanced machine translation significantly and are considered as an efficient AI tool capable of understanding spoken words almost as easily as a human can. Not only has it achieved outstanding precisional accuracy in machine learning modelling technique, it has also shown outstanding ability which has also welcomed scientists from various curriculum areas. Presently it is used as a template for critical decisions in disciplines such as healthcare, manufacturing, finance, and beyond.

Deep learning in specific has had a positive influence in traditionally challenging areas of mechanical learning [13]:

- Classification of images on a near-human level
- Recognition of speech at the nearest human level
- Transcription of handwriting at a near-human level
- Self-driven cars improved
- Digital assistant methods like Google Now, Apple's Siri, Amazon Alexa, and Microsoft Cortana further improved
- Advertisement targets improved that have been used by Google, Bing, and Baidu

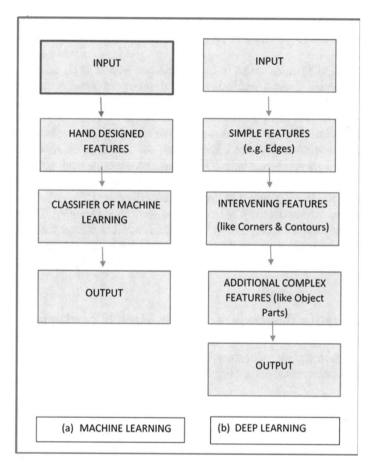

Fig. 2 (**a**) Conventional machine learning approaches which use hand-designed feature extraction algorithms. (**b**) Deep learning approach which uses a hierarchy of representations that are learnt automatically [12]

- Online searching results improved
- Ability to answer questions in a simplified natural language

Among those in which deep learning-oriented approaches are frequently employed is medical diagnosis as a critical research area [14]. Accordingly, the purpose of the chapter is to highlight recent advanced deep learning applications for the diagnosis of chronic myeloid leukaemia. Similarly, by implementing the program, someone could retrieve the relevant part of the expression data, thus allowing the classification of different subsets of genes that seem to be helpful to physicians and biologists, who have the ability to advise on clinical approaches.

DL algorithms have already exhibited advanced level efficiency in detecting the variants of cancer and have demonstrated superior accuracy compared to past functionality-engineered histology analysis methods [15].

Deep learning involves a series of prediction models which have been used recently to achieve remarkable progress in how software retrieves image features [16].

Extraction of features seems to be a significant step in implementing machine learning. Specific feature extraction method processes various cancer variants that were explored, which further had weaknesses. Representation learning was proposed in [17, 18] to resolve these shortcomings and increase the efficiency and performance. Deep learning has always had the advantage of establishing the high-level representation of features directly from the raw images. Aside from deep learning, graphics processing units (GPUs) have been used in conjunction with both feature extraction and image recognition. For instance, deep convolutional neural networks could detect cancer with accuracy and efficiency [19].

Therefore, the detection of CML requires deep machine learning. The study proposes new approach for diagnosing CML which requires a large training dataset. The significant available CML data source used here is BioGPS, from where the training dataset was collected so as to explore machine learning algorithms.

3   Progressive Diagnosis of Chronic Myeloid Leukaemia

The steps taken further on account of diagnosing chronic myeloid leukaemia are progressive as described under (Fig. 3):

3.1   Classification Algorithms Used in CML Diagnosis

Support Vector Machines (SVMs)

Support Vector Machine is a binary optimization technique that can be used for the classification of blood images sample of lymphoid stem cells and myeloid stem cells that classify chronic myeloid leukaemia and chronic lymphocytic leukaemia. Using SVM, the author achieved 92% precisional accuracy [21]. In Support Vector Machine, the input field of a dataset is separated by a different method, based on the separation line between the groups having maximum gaps. The support vectors are further classified by an algorithm called the training classifier algorithm. All core concepts, sample form, and textured characteristics are collected and reported. The most important characteristics are chosen from all apps and used for SVM preparation. The amount of lobes of nuclei, proportion of nuclei to cell, percentage of nuclei, and entropy perimeter were indeed important features preferred so far.

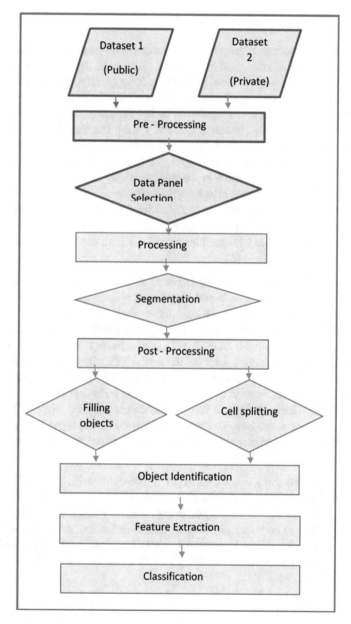

Fig. 3 Steps for progressive diagnosis of chronic myeloid leukaemia [20]

K-Nearest Neighbour (k-NN) Algorithm

The algorithm K-nearest neighbour is used to characterize cells of leukaemia from blood cells that are normal, and it has been discovered that k-NN is an all-time strong scalability classification tool [22]. The k-NN algorithm categorizes new artefacts according to similarity assessments, although it is believed that similar issues do happen in the immediate vicinity by not taking the risk of designing a model. The k-NN algorithm fits many parameters and is a slow learner without training and classification phase, or divergence, by simply memorizing the dataset used for processing. k-NN was also used to identify bursts in cells of leukaemia and the same cells were categorized with 80% precisional accuracy into chronic myelogenous leukaemia (CML) and chronic lymphocytic leukaemia (CLL). There were 12 key characteristics that were released from blood images, which were considered leukaemic to reflect the size, colour, and form, and several k-values and distance metrics were evaluated. The k-NN classifier provided strong results using a value of $k = 4$ and cosine distance metric [23].

Neural Networks Neural networks are used as it classifies the photos of blood smear among normal blood cells and blood cells of leukaemia, further helping to diagnose leukaemia using a clinical decision system. The steps in the neural networks include pre-processing, clustering, and segmenting of images. Following that, the principal component analysis (PCA) was indeed developed to obtain the key components that somehow become the input to the classifier of the neural network [24]. This classifier's aim is to do categorization between normal cells also termed as the non-cancerous cells and the abnormal cells that are the cancerous or leukaemic cells. The input layer's first two nodes are supplied with the PCA's first two main component outputs. The input nodes transfer the input load along with some weights to the hidden nodes and then the final output node value is calculated (Fig. 4).

Standard Neural Network classifier is a two-step artificial neural network, wherein the second network is used as the data to the cancer cells, and help them identify as ALL, AML, CLL, and CML using the same architecture. The work done so far also seems to be using an artificial neural network also termed as ANN to categorize blood image data into normal, AML, CML, ALL, and CLL and shows that classification performance and the accuracy could be enhanced by increasing the data size [26].

Naïve Bayes

Naïve Bayes classifier theorem had been used for the classification of leukocytes. Bayes theorem is quick and easy probability-based classifier of naïve hypothesis, which specifies the value so obtained from the functions, irrespective of the existence and non-existence of some other characteristics and of those characteristics, the likelihood of which existence contributes independently. Because it involves

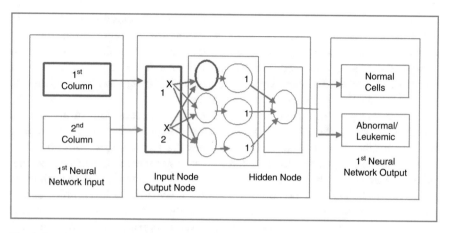

Fig. 4 Standard neural network classifier [25]

only a limited set of training data, it may approximate different parameters needed to be classified. Being a managing learning classifier, approximation of the parameters is based on the probability scheme. The mean and the variance of each and every feature of every discriminate class Cj were determined.

The independent variable is the presumed cause, which is stable and unaffected by other variables that are tried to be measured, whereupon the probability of prior $P(c)$ was recorded as

$$P(Cj) = \frac{Tcj}{Ti}$$

where Tcj is regarded as the trained images of class c and Ti is initialized as the total number of trained images, where $j = 1,2,3,4,5$, further classified as the leukocyte class [27].

Deep Learning

Deep learning has been used to classify images of the blood into CML subtypes, which are L1, L2, and L3 or regular types. CNN also called convolutional neural network has been considered as one of the deep learning algorithms commonly used in data processing of biological images. This prevents the extraction of manual features, and features are identified directly from those in the image and then transform it to an objective result with input data for the classification. To some degree, the classification process relies on the discriminating features, since the classifier is confused with too many features and too few features also aren't appropriate for successful classification.

Images of the blood cells in both patients with ALL experience and normal humans were collected as a dataset which is further divided into training and test results. It used as usual the AlexNet model with CNN for classifying ALL into its subtypes. The last three layers of the model, configured according to the data, are connected directly, softmax, and classification layer [28]. The CNN architecture is characterized as the first step, and the input layer and the convolution layer are specified afterwards. The input layer specifies the CNN image size corresponding to the image's width, height, and number of pixels.

The second layer also called the convolutional layer comprises of the neurons which interconnect the image or layer output sub-region before it. After scanning the picture, the convolutional layer determines the features specialized by those regions. There is a layer which is called a layer of normalization between the convolutional layer and the ReLU layer to boost up the training cycle and minimize sensitivity [28]. The max pooling layer fits the convolutional layer which can be used for reduction in overfitting and downsampling. In the last fully connected layer, all features are merged and the output layer comprising of the softmax function [28] is applied.

A dataset of 100 images of L1, 100 images of L2, and 100 images of L3 and 100 normal images was used, considering the training data as 80% and test data as 20% upon which the convolutional neural networks are revealed to have a precisional accuracy of 97.78% (Tables 1 and 2).

4 Artificial Neural Networks

Regardless of the nonlinear movement of neurons, neural networks are able to accomplish the task of complex computation. Artificial neural network can be used for medical images as it has the predictive capacity. The neurons are given test images for training in a general artificial neural network. An algorithm termed as the back-propagation algorithm, which is used to train neurons, having flow in the direction towards the obtained output is then compared to the desired output of the outcome, and the error signal is produced in the event that both the outputs do not match. This error of not matching propagates backwards. Weights are modified to reduce error. Repeat this step until the error comes to be zero. In the neural network there is a layering structure which includes the number of nodes that are interconnected and an activation function. Such activation functions are hyperbolic tangent function, linear function in a piece-wise manner, sigmoid function, and threshold function. Input patterns are transmitted over an input layer to the network, which further corresponds to the hidden layer, and this hidden layer further corresponds to the output layer as shown in Fig. 5.

Table 1 Comparison of classification algorithms [29]

Classification algorithm	Unsupervised/ supervised	Classes supported in nos.	Form of datasets	Datatype (linear/ non-linear)	Precisional accuracy attained
Support vector machines	Supervised	2	Small	Both linear and non-linear	92%
K-nearest neighbour	Unsupervised	>2	Small	Non-linear	80%
Neural networks	Both	>2	Small and big	Non-linear	93.7%
Naïve Bayes	Supervised	>2	Small and big	Linear	80.88%
Deep learning	Both	>2	Big	Non-linear	97.78%

Table 2 Comparison between merits and demerits of classification algorithms [30]

Classification algorithms	Merits	Demerits
Support vector machines	High precision	Requires handful memory
	Linear classification function space is not essential	Only binary classification used
	Great interaction with structured as well as semi-structured data	Not applied to large datasets
	Balances the data well to large dimensions	Training time is time consuming
K-nearest neighbour	Easy to install	Cannot manage massive datasets
	Does not demand training period	Cannot accommodate large dimensions
	Can add new data without any pause or termination	Costing of computation is very high
		Requires scaling function
Neural networks	Works well enough for both large and small datasets	Particularly susceptible to overmolding
	Requires little statistical training	Unexplained network behaviour causing problems
	Dynamic non-linear relationships between dependent and independent variables can be identified	Network length unknown
	Tolerant to failures	Works on statistical data
Naïve Bayes	Classifiers used are simple	
	Fast conversion	All attributes are assumed to be linearly independent but not in real life
	Requires less training data	Opportunity to lose accuracy and precision

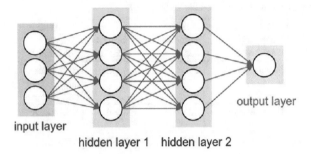

input layer

hidden layer 1   hidden layer 2

output layer

Fig. 5 Artificial neural network [31]

5   Convolutional Neural Networks

The convolutional neural network sometimes defined as ConvNet or CNN is a deep learning experience method which comprises of various counts of layers. ConvNets take inspiration from the visual biological cortex. The visual cortex includes small cell areas that are responsive to different regions of the field of vision [32]. Specific brain neurons react to various apps. Some neurons, for example, only fire in the presence of lines, some neurons fire when exposed to vertical edges, and some are of a certain orientation where horizontal or diagonal edges are seen. That concept of each of these neurons getting a ConvNets' basic role forms the basis for this.

ConvNets demonstrated excellent results on a variety of applications, including classification of images, identification of objects, speech recognition, processing of natural languages, and analysis of the medical image. Convolutional neural networks are the driving force behind computer vision, which has many applications including self-driven cars, robots, and visually impaired treatments. ConvNets' key concept is to get local patterns at higher layers from the input (typically a picture) and integrate them into further complicated components at the lower layers. It is perhaps conceptually excessive due to its own multidimensional architectural styles, and training such providers on a huge dataset appears to take a few days. And normally this deep neural network is trained on GPUs. Convolutional neural networks are so effective on human-being detection, that almost all traditional approaches are performed above [33].

5.1   Architecture of CNN

Neurons are entirely interconnected between various layers within a conventional neural network. Hidden layers are called layers which exist in between the layer called the input layer and the layer called the output layer. Every layer which is hidden consists of a number of neurons, where every other neuron in the preceding layer is completely connected to all the neurons. The problem with the neural network that is completely connected is that its network architecture which is interconnected very densely doesn't scale well to large photos. Another most effective option for large images would be to use the convolutional neural network.

ConvNets have three main characteristic features: local receptive field, weight sharing, and (pooling) subsampling.

1. *Local Receptive Field*: Though traditional neural networks include localized field architecture which is receptive in nature, i.e. every hidden unit could only relate to a specific input region called the local receptive field. It is done by raising the weight matrix or a filter to less than the input. Neurons can remove basic visual features, such as corners, edges, end points, etc., with local receptive area.
2. *Weight Sharing*: Weight sharing pertains to many regions of interest in a layer using the same filter/weights. In ConvNet, although the filters are relatively smaller than the input, each filter should be inclined at every input position, i.e. relatively similar filter is applied to all receptive local fields.
3. *Subsampling (Pooling)*: Subsampling tends to reduce input spatial size, thereby reducing the overall parameter values. There have been few subsampling technologies available and max-pooling is perhaps the most widely recognized subsampling technique.

ConvNet comprises a series of various layer types to attain different responsibilities. A neural convolutional network comprises of the following hierarchy:

- Firstly, the convolutional layer
- Secondly, the activation function layer (ReLU)
- Thirdly, the pooling layer
- Fourthly, the dully connected layer
- Finally, the dropout layer

6 Typical Application Scenario of Chronic Myeloid Leukaemia

Computer intelligence is considered to be the crucial part of smart manufacturing, recognizing that reliable insights help consider better decision-making. Machine learning has been investigated extensively at various stages of Application scenario helping in manufacturing of Cognitive Evolution, design [34], assessment, production, operation and sustainability [35] as seen. Data mining applications in manufacturing engineering are reviewed [36] that cover various categories of computational intelligence like production processes, operations, fault detection, maintenance, decision support, and improvement of product quality. Manufacturing evolution and future are analysed in [37, 38], underlining the significance of data modelling and interpretation in manufacturing computational intelligence. The

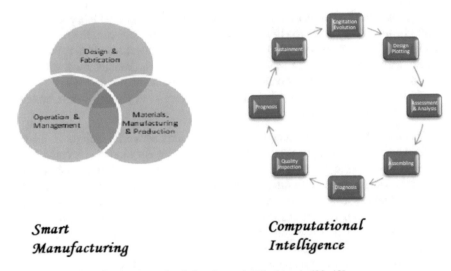

Smart
Manufacturing

Computational
Intelligence

Fig. 6 Typical application scenario of chronic myeloid leukaemia [39, 40]

machine learning application schemes in fabrication are identified as summarized in [39, 40] (Fig. 6).

Smart manufacturing also needs the capabilities of prognostics and health management (PHM) to fulfil the current and potential criteria for effective and reconfigurable production [41]. Recently, deep learning has been studied as an emerging technology for a wide variety of fabrication systems. Analysing the use of the deep learning techniques, this research deals with manufacturing, in particular in the areas of product quality testing, fault detection, and defect prognosis.

7 Conclusion

Deep learning has significantly advanced science and has made considerable advancements in our everyday lives. Many studies have adopted deep learning in the medical field and have demonstrated many notable accomplishments. One benefit towards using deep learning for training a model is its ability to start training when there are more and more data available [42]. Moreover, because health-care data have various formats, e.g. genomic data, clinical (structured) data, text and image (unstructured) data, and expression data, utilizing diverse neural network architectures to overcome various categories of data issues becomes increasingly common and useful. We have outlined in this section the recent studies that have applied deep learning methods in cancer diagnosis research. Many of those studies have demonstrated deep learning models implemented in the same or better way than other machine learning approaches [43]. Future trends will continue to concentrate on testing and refining the algorithm and developing cutting-edge models to improve the diagnosis of CML.

Key Points

1. Deep neural network (DNN) learning models embrace vast quantities of data in various formats. This platform is very useful in the diagnosis of CML because data on the health of patients include multi-source data.
2. Using extraction function may be a way to accurately retrieve data from multi-omics data used for training neural networks and potentially improving diagnosis of CML.
3. Fully linked NN and CNN models have indeed been tested to detect CML and demonstrated pretty good results in a number of studies.
4. In the CML diagnosis, existing deep learning models also need further testing and evaluation in massive datasets.

References

1. V.S.P.K. Sankara Aditya, A.B. Das, U. Saxena, Recent advances in biosensor development for the detection of cancer biomarkers. Biosen. Bioelect. **91**, 15–23 (2017)
2. U. Abbasi, P. Chowdhury, S. Subramaniam et al. A cartridge based Point-of-Care device for complete blood count. Sci. Rep. **9**, 18583 (2019). https://doi.org/10.1038/s41598-019-54006-3
3. B. Jeremiah, I. Peter, O. Anthony, A decision trees-based classification model for the survival of chronic myeloid leukaemia (cml) patients. https://www.researchgate.net/publication/309379975_A_DECISION_TREES-BASED_CLASSIFICATION_MODEL_FOR_THE_SURVIVAL_OF_CHRONIC_MYELOID_LEUKAEMIA_CML_PATIENTS. (2016)
4. M. Fruscella, A. Ponzetto, A. Crema, G. Carloni, The extraordinary progress in very early cancer diagnosis and personalized therapy: The role of oncomarkers and nanotechnology. J. Nanotech. **2016**, 18 (2016). https://doi.org/10.1155/2016/3020361
5. K.Y.P.S. Avelino, R.R. Silva, A.G. da Silva Junior, M.D.L. Oliveira, C.A.S. Andrade, Smart applications of bionanosensors for BCR/ABL fusion gene detection in leukemia. J. King Saud University – Science **29**(4):413–423 (2017)
6. M. Samandari, M.G. Julia, A. Rice, A. Chronopoulos, A.E. Del Rio Hernandez, Liquid biopsies for management of pancreatic cancer. Transl. Res. 201:98–127 (2018). https://doi.org/10.1016/j.trsl.2018.07.008. Epub 2018 Jul 26. PMID: 30118658
7. P.U. Alves, R. Vinhas, A.R. Fernandes, S.Z. Birol, L. Trabzon, I. Bernacka-Wojcik, R. Igreja, P. Lopes, P.V. Baptista, H. Águas, E. Fortunato, R. Martins, Multifunctional microfluidic chip for optical nanoprobe based RNA detection – application to Chronic Myeloid Leukemia. Sci. Rep. **8**(1):381 (2018). https://doi.org/10.1038/s41598-017-18725-9. PMID: 29321602; PMCID: PMC5762653
8. W. Jinjiang, M. Yulin, Z. Laibin, G. Robert, W. Dazhong, Deep Learning for Smart Manufacturing: Methods and Applications. Journal of Manufacturing Systems. **48**. 144–156. https://doi.org/10.1016/j.jmsy.2018.01.00 https://www.researchgate.net/publication/322325843_Deep_Learning_for_Smart_Manufacturing_Methods_and_Applications3. (2018)
9. T.H. Kim, Y. Wang, C.R. Oliver, D.H. Thamm, L. Cooling, C. Paoletti, K.J. Smith, S. Nagrath, D.F. Hayes, A temporary indwelling intravascular aphaeretic system for in vivo enrichment of circulating tumor cells. Nat. Commun. **10**(1):1478 (2019). https://doi.org/10.1038/s41467-019-09439-9. PMID: 30932020; PMCID: PMC6443676

10. E.M. Ghafoorian, Z. Fruhideh, V.M. Alviri, M.M. Asem, Electrochemical Biosensors for Cancer Detection Using Different Biomarkers. IEEE 9th Annual Computing and Communication Workshop and Conference (CCWC) **2019**, 989–996 (2019)
11. K. Doi, Computer-aided diagnosis in medical imaging: historical review, current status and future potential. Comput. Med. Imaging Graph. **31**, 198–211 (2007) [Google Scholar] [CrossRef] [PubMed]
12. U. Kose, J.A. Alzubi, Deep learning for cancer diagnosis, https://www.springer.com/gp/boo k/9789811563201#aboutBook (2021)
13. M.A. Wani, F.A. Bhat, S. Afzal, A.I. Khan, Advances in deep learning. Studies in Big Data (2020). https://doi.org/10.1007/978-981-13-6794-6
14. U. Kose, J.A. Alzubi, Deep learning for cancer diagnosis, (2021), https://www.springer.com/gp/book/9789811563201#aboutBook
15. C. Szegedy, et al., Going deeper with convolutions. *2015 IEEE Conference on Computer Vision and Pattern Recognition (CVPR)*, pp. 1–9 (2015)
16. K. He, et al., Deep residual learning for image recognition. *IEEE Conference on Computer Vision and Pattern Recognition (CVPR)*, pp. 770–778 (2016)
17. Y. Bengio, A. Courville, P. Vinvent, Representation learning: a review and new perspectives. IEEE Trans. Pattern Anal. Mach. Intell **35**, 1798–1828 (2013) [Google Scholar] [CrossRef] [PubMed]
18. Y. LeCun, Y. Bengio, G. Hinton, Deep learning. Nature **521**, 436–444 (2015) [Google Scholar] [CrossRef] [PubMed]
19. B. Xu, N. Wang, T. Chen, M. Li, Empirical evaluation of rectified activations in convolutional network. arXiv 2015, arXiv:1505.00853. [Google Scholar]
20. https://www.researchgate.net/figure/Different-steps-in-image-analysis-process_ fig4_46140875
21. J. Laosai, K. Chamnongthai, Acute and Chronic leukemia classification by using SVM and K-Means clustering. *Proceedings of the International Electrical Engineering Congress*, pp. 1–4 (2014)
22. Subhan, Ms. Parminder Kaur, Significant analysis of leukemic cells extraction and detection using KNN and Hough transform algorithm. Int. J. Comput. Sci. Trends Technol. **3**(1), 27–33 (2015)
23. N.Z. Supardi, M.Y. Mashor, N.H. Harun, F.A. Bakri, R. Hassan, Classification of blasts in acute and chronic leukemia blood samples using k-nearest neighbour. *IEEE 8th International Colloquium on Signal Processing and Its Applications*, pp. 461–65 (2012)
24. I. Vincent, K.-R. Kwon, S.-H. Lee, K.-S. Moon, Acute lymphoid leukemia classification using two-step neural network classifier. *21st Korea-Japan Joint Workshop on Frontiers of Computer Vision (FCV)*, Jan. 2015 (2015)
25. https://www.researchgate.net/figure/An-example-of-a-deep-neural-network-with-two-hidden-layers-The-first-layer-is-the-input_fig6_299474560
26. M. Adjouadi, M. Ayala, M. Cabrerizo, et al., Classification of leukemia blood samples using neural networks. Ann. Biomed. Eng. **38**(4), 1473–1482 (Apr.2010)
27. A. Gautam, P. Singh, B. Raman, H. Bhadauria, Automatic classification of leukocytes using morphological features and Naïve Bayes classifier. *IEEE Region 10 Conference (TENCON)*, pp. 1023–27, Nov. 2016
28. A. Rehman, N. Abbas, T. Saba, S.I. ur Rahman, Z. Mehmood, H. Kolivand, Classification of acute lymphoblastic leukemia using deep learning. Microsc. Res. Tech. **81**(11), 1310–1317 (2018)
29. https://www.researchgate.net/publication/305827311_COMPARATIVE_STUDY_OF_ CLASSIFICATION_ALGORITHMS_HOLDOUTS_AS_ACCURACY_ESTIMATION
30. I.J. Maria, T. Devi, D. Ravi, Machine learning algorithms for diagnosis of leukemia. Int. J. Sci. Technol. Res. **9**(01) (2020)
31. https://towardsdatascience.com/coding-neural-network-forward-propagation-and-backpropagtion-ccf8cf369f76

32. Y. Lécun, L. Bottou, Y. Bengio, P. Haffner, Gradient-based learning applied to document recognition. Proc. IEEE **86**(11), 2278–2324 (1998)
33. T. Ince, S. Kiranyaz, L. Eren, M. Askar, M. Gabbouj, Real-time motor fault detection by 1-D convolution neural networks. IEEE Trans. Ind. Electron. **63**(11), 7067–7075 (2016)
34. W. Zhang, M.P. Jia, L. Zhu, X. Yan, Comprehensive overview on computational intelligence techniques for machinery condition monitoring and fault diagnosis. Chin. J. Mech. Eng. **30**(4), 1–14 (2017)
35. J. Lee, E. Lapira, B. Bagheri, H. Kao, Recent advances and trends in predictive manufacturing systems in big data environment. Manuf. Lettersm. **1**(1), 38–41 (2013)
36. J.A. Harding, M. Shahbaz, A. Srinivas Kusiak, Data mining in manufacturing: a review. J. Manuf. Sci. Eng. **128**, 969–976 (2006)
37. B. Esmaeilian, S. Behdad, B. Wang, The evolution and future of manufacturing: a review. J. Manuf. Syst. **39**, 79–100 (2016)
38. H.S. Kang, Y.L. Ju, S.S. Choi, H. Kim, J.H. Park, Smart manufacturing: past research, present findings, and future directions. Int. J. Precision Eng. Manuf. Green Technol. **3**(1), 111–128 (2016)
39. B.T. Hazen, C.A. Boone, J.D. Ezell, L.A. Jones-Farmer, Data quality for data science, predictive analytics, and big data in supply chain management: an introduction to the problem and suggestions for research and applications. Int. J. Prod. Econ. **154**(4), 72–80 (2014)
40. S.J. Shin, J. Woo, S. Rachuri, Predictive analytics model for power consumption in manufacturing. Procedia CIRP **15**, 153–158 (2014)
41. G.W. Vogl, B.A. Weiss, M. Helu, A review of diagnostic and prognostic capabilities and best practice for manufacturing. J. Intell. Manuf., 1–17 (2016)
42. A. Esteva, A. Robicquet, B. Ramsundar, V. Kuleshov, M. DePristo, K. Chou, C. Cui, G. Corrado, S. Thrun, J. Dean, A guide to deep learning in healthcare. Nat. Med. **25**, 24–29 (2019) [CrossRef] [PubMed]
43. A.A. Elfiky, M.J. Pany, R.B. Parikh, Z. Obermeyer, Development and application of a machine learning approach to assess short-term mortality risk among patients with cancer starting chemotherapy. JAMA Netw. Open **1** (2018) [CrossRef]

An Automatic Bean Classification System Based on Visual Features to Assist the Seed Breeding Process

Miguel Garcia ⓘ, Deisy Chaves ⓘ, and Maria Trujillo ⓘ

1 Introduction

The food industry plays a crucial role in the global economy. zMost countries, particularly in Africa and Latin America, have food products as part of their main exportations. Nowadays in the competitive market, producers have to sell products which are sorted according to physical characteristics, such as appearance, size, colour, internal health, and variety, since people tend to consume healthy and quality products. Besides, identifying optimal grain varieties helps farmers use suitable grains for planting and marketing since marketers have essential standards for grains [1].

Common bean (*Phaseolus vulgaris* L.) is one of the most important grain legumes worldwide for human consumption because it is a low-cost source of dietary proteins [2]. In Latin America, the growth of bean production has been supported during the last decades by a sustained work of plant breeding programmes, resulting in the development of a large number of bean varieties [2]. In particular, the International Center for Tropical Agriculture (CIAT) in Cali, Colombia, has a programme that investigates and develops varieties of beans with genetic resistance to the main pests and diseases that affect crops. The CIAT genetic improvement consists of hybridisations between bean breeding lines to select the best offspring, aiming to improve the qualities of the parent lines and generate new varieties adapted to drought stress conditions and resist different diseases, such as

M. Garcia (✉) · M. Trujillo
Multimedia and Computer Vision Group, Universidad del Valle, Cali, Colombia
e-mail: luis.ampudia@correounivalle.edu.co; maria.trujillo@correounivalle.edu.co

D. Chaves
Multimedia and Computer Vision Group, Universidad del Valle, Cali, Colombia

Group for Vision and Intelligent Systems, Universidad del León, León, Spain
e-mail: deisy.chaves@correounivalle.edu.co

© Springer Nature Switzerland AG 2022
P. Johri et al. (eds.), *Trends and Advancements of Image Processing and its Applications*, EAI/Springer Innovations in Communication and Computing,
https://doi.org/10.1007/978-3-030-75945-2_8

Table 1 Phenotypic characteristics used to describe beans during the manual inspection [2]

Feature	Feature values
Dominant and secondary colour	White, cream-beige, yellow, brown-maroon, pink, red, purple, black, others
Size	Seeds up to 25 g (small), seeds between 25 g and 40 g (medium), seeds larger than 40 g (large)
Shape	Oval, round, and elongated
Brightness	Bright or opaque

anthracnose, brown spot, angular leaf spot, and bean common mosaic virus [3]. During this process, researchers/breeders inspected bean seeds manually to determine which varieties may be discarded and which ones continue to the next line of breeding by considering the phenotypic characteristics described in Table 1, such as non-striking colour and non-common shape and size, among others. Thus, the progress of breeding programmes depends on the precise phenotypic characterisation of beans with new or improved features. However, bean seeds have a wide variance of colour, shape, size, and brightness due to the genetic diversity that exists within bean species making the phenotypic characterisation highly subjective, error-prone, and time-consuming.

Nowadays, computer vision techniques appear as one of the key assets to foster the food industry and increase competitiveness by providing an objective way to measure relevant visual features related to food quality, such as shape, size, or colour [4]. These techniques have been used to automate the quantification of bean phenotypic features using digital images and their classification into bean varieties [1, 4–9]. However, most of the approaches for automatic bean classification rely on licensed software to analyse the images, as Matlab [1, 4, 7] or KS-400 [5, 6], and do not consider the effect of "glued" seeds in the input images [1, 4–7] which may affect the measurement of shape and size characteristics commonly used in the bean inspection. Therefore, automation of bean classification is still an open problem.

In this paper, we propose an automatic three-fold system to classify bean seeds through computer vision and supervised machine learning algorithms implemented with open-source libraries. First, bean images are taken using an image acquisition protocol designed to improve the contrast between the background and bean seeds. Second, seeds are segmented using a combination of thresholding and watershed [10] methods aiming to identify correctly "glued" seeds. Third, individual seeds are classified using a classification model built with phenotype and morphology features. Besides, we validated our system using images from six of the most used bean varieties for breeding at CIAT (see Fig. 1): MBC 46, Bola Roja, CAL 96, Cargamanto Rojo, Cargamanto, and MAC 31. Some of the selected bean varieties present visual similarities in colour, shape, and size, making its classification difficult. Results show that the proposed system accurately classified the bean varieties, and it is suitable for supporting the inspection of beans during seed breeding.

Fig. 1 Selected bean varieties for automatic classification

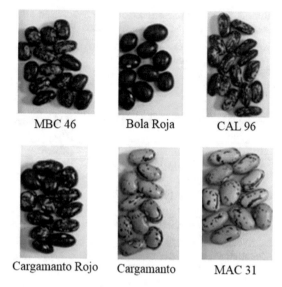

MBC 46   Bola Roja   CAL 96

Cargamanto Rojo   Cargamanto   MAC 31

2 Related Works

In the literature, the automatic classification of bean seeds has been addressed using machine learning algorithms with visual features obtained from bean seed images, such as colour, shape, and size [1, 4–9]. Table 2 presents a summary of the reviewed works.

In [4], a system was developed based on the skin colours of the grains for classifying images of five bean classes. The system employed edge detection, mathematical morphology operators, and colour features quantified by statistical moments using the Matlab software. A multilayer perceptron is used for the classification task obtaining a success rate of 90.6%. Similarly, a system was proposed in [1] using a multilayer perceptron neural network and colour characteristics to classify images of ten bean varieties. Experiments showed that the system achieved a sensitivity of 96% and a specificity of 97.1%.

In [5], a system was proposed based on linear discriminant analysis for classifying images of six landraces of beans from Italy. The experiments evaluated size, shape, colour, and texture features extracted from grains using the image analysis library KS-400. The system achieved a success rate of 99.56%. In a subsequent work [6], the same authors conducted new experiments considering 15 traditional Italian landraces of beans and obtained a success rate of 98.49%.

In [7], a system was developed to distinguish seven dry beans varieties with similar colour in a digital image. Twelve-dimensional and four shape features were obtained from the grains using the Matlab software to build four classification models: a multilayer perceptron, support vector machine (SVM), k-nearest neighbours, and decision tree. Results show that SVM achieved the highest accuracy (93.13%).

Table 2 Summary of bean seed classification systems based on image analysis. Higher accuracy (Acc.) values indicate a better performance

Year and reference	# of classified bean varieties	Similar beans	Classification features	Classification method	Acc. (%)
2007, [4]	5	No	Colour and size	Neural networks	90.60
2007, [5]	6 Italian beans	–	Colour, size, texture, and shape	Linear discriminant analysis	99.56
2009, [6]	15 Italian beans	Yes	Colour, size, and shape	Linear discriminant analysis	98.49
2013, [1]	10	Yes	Colour	Neural networks	N/A[a]
2015, [8]	3 Brazilian beans	No	Colour	k-nearest neighbours	99.88
2016, [9]	3 Brazilian beans	No	Colour	Neural networks	99.14
2020, [7]	7 dry beans	Yes	Dimensional and shape forms	SVM	93.13

[a]The study [1] reported as evaluation metrics a sensitivity of 96% and a specificity of 97.1%

In [8], a robust correlation-based granulometry module was proposed for segmenting seeds of three Brazilian beans and classifying them using colour features with the k-nearest neighbour algorithm. The system achieved a success rate of 99.88%. This study was extended in [9] to improve the processing time by using the watershed transform along with the refinement heuristics to segment grains and a multilayer perceptron neural network to classify them. The system was evaluated using the same Brazilian bean varieties of [8] and yielded a success rate of 99.14%.

During the literature review [1, 4–9], we observed that colour, shape, and size features are the most appropriate for automatic bean characterisation and supervised learning algorithms. Most of these approaches used software licensed for processing bean images, such as Matlab [1, 4, 7] or KS-400 [5, 6]. Moreover, few proposals [8, 9] consider the effect of "glued" seeds during the identification of individual beans before their classification, although the incorrect segmentation of "glued" seeds may affect the measurement of shape and size features and the bean classification. To cope with these drawbacks, our bean classification system was developed using open-source libraries and incorporated a segmentation method robust to "glued" seeds that combine thresholding and watershed methods.

3 Proposed Approach

Figure 2 illustrates the three components of the proposed bean classification system. First, bean images are acquired following a protocol designed to reduce shadows and improve the contrast between the background and bean seeds. Second, individual seeds are obtained from the acquired images using a combination of Otsu [11] and the watershed [10] segmentation methods. Third, seeds are described using phenotype features – i.e. colour, brightness, area, perimeter, and maximum and

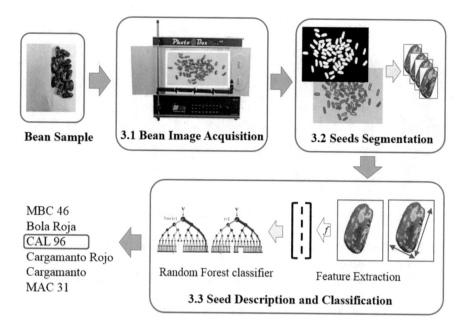

Fig. 2 Proposed bean classification system

minimum diameters – and classified with a random forest model into six bean varieties: MBC 46, Bola Roja, CAL 96, Cargamanto Rojo, Cargamanto, and MAC 31.

3.1 Bean Image Acquisition

The performance of a classification system heavily depends on the image acquisition process. Thus, we proposed a ten-step acquisition protocol to homogenise the capture of bean images focused on improving the contrast between the background and bean seeds, as well as to reduce the presence of shadows that make the segmentation and description of beans difficult. The image acquisition protocol is described next.

1. Place a Nikon D3300 camera on the tripod at the centre of a studio Photo-eBox Plus (PeP) with a height of 30 cm. We selected the PeP because it has multiple switches that allow on-off control over fluorescent lights around the box.
2. Place a blue sample holder of 10 × 13 cm inside the PeP. We choose this colour as a background because it improves the contrast with beans [1].
3. Turn on the lights Left-Side Light and Right-Side Light in the PeP.
4. Start the Camera Control Pro program.
5. Select the Exposure 1 tab, then set the Shutter Speed to 1/50 sec, the Aperture to f/16, and Exposure Mode to Manual.

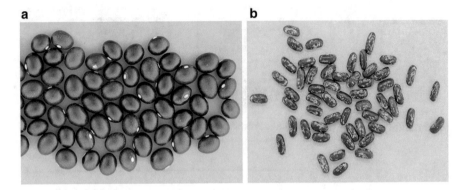

Fig. 3 Image taken (**a**) without the proposed acquisition protocol and (**b**) following it

6. Select the Exposure 2 tab, then set the ISO Sensitivity to ISO 500 and the White Balance to Auto.
7. Select the Storage tab, then set the Data Format to JPEG (Fine) and the Image Size to 4288×2848 pixels.
8. Select the Download Options tab and select the folder where the taken images are going to be saved.
9. Load up the beans on the sample holder.
10. Take the image.

Figure 3 shows bean images taken following our acquisition protocol.

3.2 Seed Segmentation

An accurate bean seed segmentation is crucial to compute the phenotype features required to describe and classify bean images. Even though humans can easily identify individual bean seeds, several factors such as "glued" or touching beans make the automatic identification of individual seeds challenging. We proposed the next eight-step strategy for the segmentation of bean seeds (see Fig. 4).

1. RGB bean images are transformed to greyscale and filtered using a median filter with a size window of 5×5 pixels to remove noise due to light variability during image acquisition (see Fig. 4b).
2. Binary images are obtained from filtered greyscale images by applying the Otsu algorithm [11] and morphological operations of opening and closing to remove small objects and fill beans by closing small hole (Fig. 4c).
3. Regions containing beans or foreground are identified using the distance transform to detect "glued" or touching seeds. In particular, a contour is created as far as possible from the centre of the overlapping beans by thresholding the distance

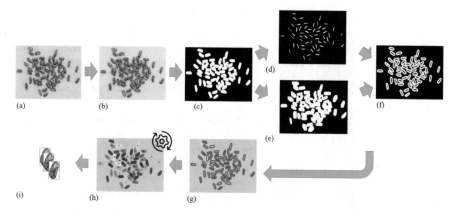

Fig. 4 The proposed strategy for the segmentation of individual bean seeds. (**a**) Input RGB image. (**b**) Greyscale image. (**c**) Binary image. (**d**) Foreground region in while. (**e**) Background region in black. (**f**) Unknown region in white, subtracting foreground from background region. (**g**) RGB mask image bulit using lablled region. (**h**) Updated mask image after watershed performs. (**i**) Segmented bean seeds (output)

transform images. In this work, a threshold value of 60, set experimentally, is used (see Fig. 4d).

4. Regions corresponding to the background are obtained by applying a dilation operation on binary images. In this study, the dilation operation is selected because it increases object boundaries, allowing the identification of background areas (see Fig. 4e).

5. Regions containing unknown areas that may correspond either to the foreground or to the background areas are obtained. Unknown areas are generally around seed boundaries where the foreground and the background meet, or two different seeds meet. The unknown regions are computed by subtracting the foreground region from the background region (see Fig. 4f).

6. An RGB mask image is created by labelling the foreground, the background, and the unknown regions using the *connectedComponents* method (see Fig. 4g).

7. The watershed method [10] is applied to the RGB mask image to identify individual seeds. After that, the *findContours* method is used to obtain each of bean seeds' contour and generated an updated RGB mask image (see Fig. 4h).

8. Individual bean seeds are extracted using the *connectedComponents* method on the updated RGB mask image (see Fig. 4i).

3.3 Bean Description and Classification

After the bean seeds are segmented, a feature vector is calculated from each seed image to build classification models, using seven phenotype features (see Table 3): predominant and secondary colour, brightness, and shape – i.e. area, perimeter, and

Table 3 Description of features used to represent bean seed images

Feature	Illustration	Description
Predominant and secondary colours		First and second colours on the RGB bean image with the higher percentages of occurrences after grouping the image colours with the K-means algorithm. In this example, the predominant and secondary colours are identified in beige and brown colour, respectively
Brightness		Average of the brightness values (B) computed at each pixel on the RGB bean image, as: $$B = \sqrt{0.241 \times r^2 + 0.691 \times g^2 + 0.068 \times b^2},$$ where r, g, and b correspond to the red, green, and blue channels of the RGB image
Area		Total number of white pixels on the binary seed image
Perimeter		Total number of white pixels on the contour of the seed image
Maximum and minimum diameters		The major and the minor axis lengths on the binary seed image

maximum and minimum diameters. We selected these features to describe the content of bean seed images based on the reported in the literature and the phenotype characteristics used during the manual bean inspection at CIAT (see Table 1).

Finally, we train a bean seed classifier using the random forest algorithm with a set of feature vectors obtained from seed images and their corresponding labels – i.e. MBC46, Bola Roja, CAL 96, Cargamanto Rojo, Cargamanto, and MAC 31. Random forest was chosen based on the results of one of our previous works [12].

Random forest is a supervised learning algorithm that builds an ensemble of decision trees by following a three-step process [13, 14]. First, T subsets are selected from the training data, i.e. bean feature vectors and labels. The parameter T corresponds to the number of trees in the ensemble. Second, a decision tree with D nodes is grown for every subset of the training data. Third, the outputs of the T trees are combined to classify the test data. Hence, random forest avoids over-fitting because individual classifiers do not use the whole training data [14].

4 Experimental Setup

In this study, a set of 600 images −100 images per evaluated bean variety; a total of 46,440 individual seeds – are used to assess the proposed system for classifying bean seed images as MBC 46, Bola Roja, CAL 96, Cargamanto Rojo, Cargamanto, and MAC 31. Our system was coded using Python 3, an open-source programming language commonly used to build computer vision applications and train machine learning models.

We used 70% of the seed images for training the random forest models and the remaining images for testing. The models were built using the phenotype features extracted from bean seed images (see Sect. 3.3) by following a fivefold cross-validation strategy to optimise the parameters T and D of the random forest model. In the cross-validation strategy, the training data is divided randomly into five subsets (the folds) of equal size; one subset is used for evaluation, and the remaining ones are used to train the classifier. This process is repeated five times (the folds) until each subset is used for evaluation, and the random forest classifier that yielded the highest performance measurement, i.e. classification accuracy, is selected for testing.

In particular, we evaluated the performance of random forest classifiers with the accuracy (Acc) and the F1-score measures [15]. The accuracy is computed as the number of bean seed images correctly classified divided by the set's cardinality. The F1-score is obtained to the harmonic mean of the precision and the recall measures. Moreover, we compare the results of our proposal against a state-of-the-art approach for bean classification based on a multilayer perceptron neural network and colour features [1].

5   Experimental Results and Discussion

Table 4 shows the accuracy value of random forest models built varying the number of trees, $T = \{50, 100, 200\}$, and the maximum number of nodes, $D = \{6, 10\}$. Results show that the best bean classifier is the model built using $T = 200$ trees with $D = 10$ nodes. This model achieved on the testing dataset an accuracy value of 98.5% and an F1-score value of 0.98, indicating that the bean classifier has a good generalisation capability.

Moreover, we analysed the performance of our classifier per bean seed variety; see Table 5. As we can observe, our model is able to classify correctly four out of six bean varieties with precision, recall, and F1-score values of 1.0. Note that the two bean varieties – Cargamanto Rojo and MBC 46 – classified with less accuracy achieved precision, recall, and F1-score values higher than 0.93 despite that both bean varieties have very similar visual features as is illustrated in Fig. 5.

Finally, we performed a comparison between the results of our bean classification system and the ones obtained using a state-of-the-art approach for classifying ten bean varieties based on a multilayer perceptron neural network with predominant and secondary colour features [1]. Since the source code for the state-of-the-art approach was not publicly available, we implemented from scratch the network architecture described in [1] adapting the output layer to six neurons that correspond to the number of bean varieties considered in this study. Accuracy values were computed for both approaches on the testing set, and the results showed that

Table 4 Performance of random forest models varying the parameters T and D. The best accuracy (Acc.) value is highlighted in bold

Parameter		
Number of trees, T	Maximum nodes, D	Acc. +/− I.C (%)
50	6	97.6 ± 0.005
100	6	97.7 ± 0.005
200	6	97.7 ± 0.005
50	10	98.3 ± 0.005
100	10	98.4 ± 0.006
200	**10**	**98.4 ± 0.005**

Table 5 The precision, the recall, and the F1-score values computed per bean seed variety using the best random forest classifier

Bean variety	Precision	Recall	F1-score
Bola Roja	1.00	1.00	1.00
CAL 96	1.00	1.00	1.00
Cargamanto	1.00	1.00	1.00
Cargamanto Rojo	0.97	0.93	0.95
MAC 31	1.00	1.00	1.00
MBC 46	0.93	0.97	0.95
Avg. total	0.98	0.98	0.98

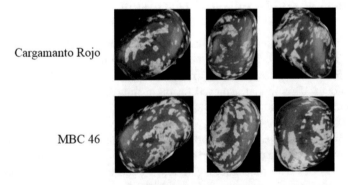

Cargamanto Rojo

MBC 46

Fig. 5 Images from bean varieties of Cargamanto Rojo and MBC 46

our proposal outperformed the approach based on the multilayer perceptron neural network (accuracy, 98.5% vs 87.3%). These results indicate that we selected the most appropriate features to correctly distinguish, in most cases, bean varieties with a similar appearance. Therefore, the proposed system may be suitable to support researchers/breeders during the inspection of bean seeds in the breeding process by accurate classification of bean seed images.

6 Conclusions

In this work, we proposed an automatic system based on computer vision and supervised learning algorithm to classify six of the most used bean varieties for breeding at CIAT: MBC 46, Bola Roja, CAL 96, Cargamanto Rojo, Cargamanto, and MAC 31. Besides, our system includes a protocol to homogenise the acquisition of bean images, a segmentation strategy based on thresholding and the watershed methods that identify successfully "glued" or touching seeds, and a random forest model to classify the identified beans based on seven phenotype features: predominant and secondary colour, brightness, and area, perimeter, and maximum and minimum diameters.

Experimental results indicate that the proposed bean classification system is highly accurate, achieving an accuracy of 98.5% and an F1-score of 0.98. Thus, it can be used to support bean inspection during the breeding process. As future work, we will extend the study by including more bean varieties.

Acknowledgements We would like to thank the bean inspectors from the CIAT for detailing the bean inspection process and helping with image acquisition.

References

1. A. Nasirahmadi, N. Behroozi-Khazaei, Identification of bean varieties according to color features using artificial neural network. Spanish J Agric Res **11**, 670–677 (2013)
2. O. Voysest, Mejoramiento genético del frijol (Phaseolus vulgaris L.): legado de variedades de América Latina, 1930–1999 (2000).
3. J. Arias, T. Rengifo, M. Jaramillo, *Manual técnico: buenas prácticas agrícolas (BPA) en la producción de frijol voluble*, 1st edn. (Print Ltda, La selva, Medellín, Colombia, 2007)
4. K. Kılıç, İ.H. Boyacı, H. Köksel, İ. Küsmenoğlu, A classification system for beans using computer vision system and artificial neural networks. J. Food Eng. **78**, 897–904 (2007)
5. G. Venora, O. Grillo, C. Ravalli, R. Cremonini, Tuscany beans landraces, on-line identification from seeds inspection by image analysis and linear discriminant analysis. Agrochimica **51**(4-5), 254–268 (2007)
6. G. Venora, O. Grillo, C. Ravalli, R. Cremonini, Identification of italian lanraces of bean (phaseolus vulgarisl.) using an image analysis system. Scientia Horticulturae **121**(4), 410–418 (2009)
7. K. Murat, A. Ilker, Multiclass classification of dry beans using computer vision and machine learning techniques. Comput. Electron. Agric. **174**, 105507 (2020)
8. S.A. De Araújo, J.H. Pessota, H.Y. Kim, Beans quality inspection using correlation-based granulometry. Eng. Appl. Artif. Intel. **40**, 84–94 (2015)
9. P.A. Belan, S. Araújo, W.A.L. Alves, An intelligent vision-based system applied to visual quality inspection of beans. In: *International Conference Image Analysis and Recog-nition*, pp. 801–809, Springer (2016).
10. Beucher, S., Lantuéjoul, C.: Use of Watersheds in Contour Detection, (1979).
11. N. Otsu, A threshold selection method from gray-level histograms. IEEE Trans. Syst. Man Cybern. **9**, 62–66 (1979)
12. M. Garcia, M. Trujillo, D. Chaves, Global and local features for bean image classification. In: *7th Latin American Conference on Networked and Electronic Media (LACNEM 2017)*, pp. 1–6, (2017).

13. A. Criminisi, J. Shotton. Decision forests for computer vision and medical image analysis. Springer Science & Business Media, Berlin (2013) https://doi.org/10.1007%2F978-1-4471-4929-3
14. F. Tang, H. Lu, T. Sun, X. Jiang, Efficient image classification using sparse coding and random forest. In: *2012 5th International Congress on Image and Signal Processing*, pp. 781–785 (2012).
15. D.M. Powers, Evaluation: from precision, recall and f-measure to ROC, informedness, markedness and correlation. J. Mach. Learn. Technol. **2**, 37–63 (2011)

Supervised Machine Learning Classification of Human Sperm Head Based on Morphological Features

Natalia V. Revollo ⓘ, G. Noelia Revollo Sarmiento ⓘ, Claudio Delrieux ⓘ, Marcela Herrera ⓘ, and Rolando González-José ⓘ

1 Introduction

Biomedical technologies are driving innovative research and development initiatives worldwide, with the general aim of improving human health. Image-based diagnosis, in particular, is triggering novel image interpretation methodologies in almost all imaging modalities related to human health. Infertility is a condition that affects approximately 48.5 million couples worldwide [1]. Semen analysis is considered the main diagnostic source of information in the evaluation of male reproductive capacity [2, 3]. This analysis is routinely performed using images taken with optic microscopy devices on carefully produced and prepared samples. The whole image interpretation procedure requires highly skilled specialists with specific training in andrology and related subjects. Notwithstanding the background and prior experience of the specialists, interpretation is commonly biased, due to intra- and inter-subjective variance [4, 5]. Many times a specific spermatozoon, taken on

N. V. Revollo (✉)
Instituto de Investigaciones en Ingeniería Eléctrica, Consejo Nacional de Investigaciones Científicas y Técnicas, Bahía Blanca, Argentina

Departamento de Ingeniería Eléctrica y Computadoras, Universidad Nacional del Sur, Bahía Blanca, Argentina
e-mail: nrevollo@criba.edu.ar

G. N. R. Sarmiento · C. Delrieux
Departamento de Ingeniería Eléctrica y Computadoras, Universidad Nacional del Sur, Bahía Blanca, Argentina

M. Herrera
VITA Medicina Reproductiva, Puerto Madryn, Chubut, Argentina

R. González-José
Instituto Patagónico de Ciencias Sociales y Humanas, Consejo Nacional de Investigaciones Científicas y Técnicas, Puerto Madryn, Chubut, Argentina

© Springer Nature Switzerland AG 2022
P. Johri et al. (eds.), *Trends and Advancements of Image Processing and its Applications*, EAI/Springer Innovations in Communication and Computing, https://doi.org/10.1007/978-3-030-75945-2_9

a standard sample, is considered healthy by some specialists and unhealthy by others. This makes human-assisted semen analysis highly variable, and therefore to obtain reliable information the usual recommendation is to repeat the analysis several times and on several samples [6–8].

On the other hand, apart from fertility-related conditions, there are increasing requirements for innovative and improved methods to analyze human semen, including contraception effectiveness assessment, concerns regarding the possible incidence of environmental contamination in the reproductive function [1], and many others. Also, there is a growing economic impact related to the improvement of sperm diagnosis and selection in animal production industries [9–11]. Human sperm features vary according to age, seasonal and environmental factors, body temperature, geographic context, and living habits [12, 13]. The basic parameters taken into account to evaluate samples are seminal volume, pH, concentration, motility, and sperm morphology, the latter being specifically relevant but difficult to assess. Specialists' training requires permanent updating through the release of periodicals and white papers. The most renowned is published by the World Health Organization (WHO) [1]. Despite the existence of these references, there is no consensus about simple objective methodological or experimental evaluation criteria [11, 14–17]. This makes the validation or the comparison of evaluations among different specialists difficult, and therefore in clinical studies where precision is essential, analyses cannot be handled in a trustworthy manner in low-to-medium complexity laboratories, and samples are sent to highly specialized centers. This is not only an expensive process but also implies risks such as uncontrolled sample deterioration during shipping and handling.

The use of alternative techniques, such as computer-aided sperm analysis (CASA), is on the rise since they perform in a reproducible way, and therefore bias can be adjusted systematically. Both semi-automated and automated methods for sperm morphology assessment have been developed in the last decades. The earliest approaches, based on direct measurements, classify human sperm morphology by means of segmentation, using a threshold that considers the difference in pixel intensity between the sperm head and the background in the sperm image [18–20]. Park et al. [21] proposed a method based on the Hough transform for sperm head segmentation. The sperm head boundary is approximated with an ellipse, and basic morphological parameters can be calculated after boundary estimation. Heads of normal sperms show little deviation between their actual boundary and the estimated elliptical boundary, while abnormal sperm heads diverge significantly. Park and Paick [22] classify sperm head boundaries according to elliptic features using a method based on Fourier shape analysis, where a dyadic wavelet transform is employed. The sperm boundary is parameterized with a discrete Fourier transform and reconstructed by this ancillary wavelet function. Carrillo et al. [23] developed a method for segmenting sperm into nucleus, acrosome, and mid-piece. The method is structured according to the following stages: detection and extraction of individual spermatozoon, image enhancement, and segmentation algorithm. This approach represents one of the first successful attempts to discriminate among sperm parts using shape analysis criteria. Abbiramy and Shanthi [24] focused on the classification of spermatozoa as either normal or abnormal. The stages of this technique involve color space conversion, image noise detection and removal, extraction of individual sperms using Sobel edge detection, and the segmentation into various regions of interest such as sperm head, mid-piece, and tail. The last stage involves a statistical measurement of sperm features. Bijar et al. [25] proposed a method for sperm segmentation into acrosome, nucleus, mid-piece, and tail. The whole tail was identified through some points placed on the sperm tail. The estimated points can be used to determine some features of the tail such as length and shape, among others. The acrosome, nucleus, and mid-piece were segmented using a Bayesian classifier and a Markov random field (MRF) model.

More recently, several authors (see for instance [26–29]) developed methods for sperm segmentation and classification. Chang et al. [26] presented an improved framework to detect and segment sperm heads. The framework uses a clustering method and image processing techniques in different color spaces. Its main contribution is a gold standard dataset which represents the state-of-the-art reference for further evaluation and comparison with future developments in the area. Shaker et al. [27] provided a method to segment sperm parts using active contours in the HSV color space, facilitating the recognition of the tail. A tail point is used to remove the mid-piece from the segmented head. Ghasemian et al. [30] presented an automatic sperm morphology analysis (SMA) method for abnormal human sperm detection. The method was used to detect and analyze sperm parts. The first stage involves noise reduction and contrast enhancement of the image. In the second stage, sperm part recognition and shape and size analysis of each part are performed. Finally, each sperm is classified as either normal or abnormal using SMA. Tseng et al. [31] presented a computer-assisted system for sperm morphology diagnosis. The system classifies sperm as normal or abnormal considering a one-dimensional waveform and gray-level features. The two main contributions of this paper are an approach to transform the sperm contour into a one-dimensional waveform used as a feature, and a classification of the waveform and gray-level features using SVM.

Other approaches consider the use of principal component (PC) and discriminant analysis to evaluate sperm morphometrics [32–38]. These techniques are used to reveal subpopulations of spermatozoa. Results showed that 13 specific features can be represented using fewer PCs, 2 in bulls, adolescent humans, adult human sperm head DNA, domestic cats, puma, roosters, and guinea fowls. Three PCs are used in adult human split ejaculate samples. From these PC variables, discriminant analysis was employed to separate homogeneous subpopulations of morphological forms. There are also available commercial CASA (computer-aided sperm analysis) software products, which intend to automate the evaluation of some of the parameters related to semen quality (see for instance [39–41]). Usually these products measure the concentration, motility, and kinematics of spermatozoa and the percentage of motile spermatozoa, among other parameters. Some systems include modules for sperm concentration estimation. However, few products include modules for automatic or semi-automatic morphological measurement. In particular, sperm morphology is seldom analyzed automatically, due to the complex image processing requirements [42, 43]. Until recently, sperm concentration measurements with

CASA systems presented bias and errors due to the confusion between particle debris and actual spermatozoa [44]. DNA fluorescent staining is the most popular technique used to solve these problems, making possible to determine the sperm structure [45, 46]. Another difficulty is the segmentation of sperm parts. A variety of algorithms for unsupervised tail segmentation have been proposed, but in general the problem still requires the application of more advanced methods. The aim of this paper is to develop an automatic framework to classify sperm head morphology using image processing and machine learning techniques.

2   Methodology

2.1   Data Acquisition

To evaluate the methodology, we used the human sperm head morphology (HuSHeM) dataset [27], gold standard dataset [26], and also images from a reproductive health institute (VITA Medicina Reproductiva, IPAMER SRL, Chubut, Argentina). Samples at the VITA institute were stained with a rapid staining procedure usually used for sperm morphology [1]. For fixing the air-dried semen smear, the slides were immersed in triarylmethane fixative for 15 seconds. The excess was drained by placing slides vertically on an absorbent material. The fixed semen smears were stained using eosinophilic xanthene. For capturing the images we used a Leica DFC 450 Camera and a Leica DM 2500 microscope. The resolution of each image is 25 Mpixels in RGB color space. In all, a total of 20 plates from the gold standard dataset and 40 plates from the VITA dataset were used to develop and test the segmentation stage. The HuSHeM dataset was also included (applying the same random transformations), and therefore a total of 2590 sperm head samples were used for training and testing the classifier.

2.2   Data Preparation

Images from different sources and staining were selected to train the automatic sperm segmentation system. For each image, both metadata (such as data source, organization, acquisition date, and stain method) and the mean, covariance matrix, and maximum and minimum values in the YIQ color space were added and stored in a knowledge database. Also, several sperm samples were manually selected, and their YIQ values were computed and stored. This information constitutes a feature vector (foreground and background color prototypes) for each image. This color space was chosen given that it splits chromaticity and luminance, and provides reasonable metric properties (i.e., similarity between two colors can be approximated using their distance in the YIQ space).

Other color spaces with similar properties like Lab, Luv, and L*u*v* were also tested, achieving similar results, but since they require more complex processing than YIQ, this color space was finally adopted [47]. The first step during the morphological classification, when a new sperm image arrives, is to recognize a reference image with similar stain. For this we used the minimum distance method in the aforementioned color space [48] to find in the knowledge database an image with the most similar stain. Once a reference image is found with the closest color statistics as the one being processed, the color prototypes for foreground and background of the reference image can be successfully applied to automatically segment the spermatozoa in the new sperm image, as will be shown in the next subsection.

2.3 Sperm Segmentation

The goal of the sperm segmentation procedure is to extract the sperm heads that appear in the image. Our study is focused only on the morphological properties of the sperm head, and therefore, we are not considering the sperm tails, which can be added to enhance the system performance in future developments. Once foreground and background color prototypes are defined, the Mahalanobis distance d of the color of each pixel p in the sperm image is computed:

$$d(\vec{p}) = \sqrt{\left(\vec{p}-\vec{b}\right)^T \sum^{-1}\left(\vec{p}-\vec{b}\right)} \tag{1}$$

where $\vec{p}$ represents a (column) vector with the YIQ color components of the pixel p, and b and Σ represent, respectively, a vector with the mean YIQ components and the covariance matrix of the background prototype of the reference image. The whole set of (scalar) distances d can be regarded as a population with two modes, a larger one comprising the background pixels (with very low d values) and a smaller mode comprising the foreground pixels (with higher d values). Finally, every pixel can be tagged as either foreground or background using a maximum likelihood estimation criterion, producing a binary mask. This process is illustrated in Fig. 1, showing a

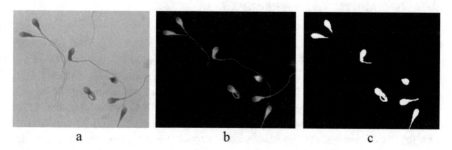

a   b   c

Fig. 1 Segmentation of sperm heads. (**a**) Sperm image, (**b**) Mahalanobis distances with respect to background color prototype (represented in grayscale, darker means closer), and (**c**) binary mask

Fig. 2 (**a, b**) Represent sperm head in the image and (**c, d**) are the distance images to the pixel prototypes.

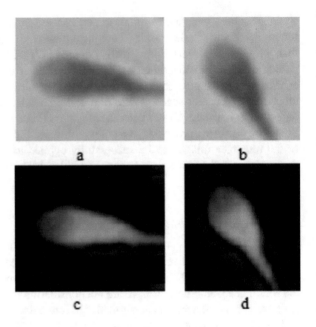

sperm image, the distance in color space of each pixel to the background prototype (or *distance image*, rendered in grayscale), and the final binary mask. Additional processing is often required, since the middle piece and initial part of tail can be stained with a color similar to the head (see Fig. 2). To take out the pixels which are not part of the head, we apply a Gaussian filter with $\theta = 2.0$ over the binary mask (see Fig. 1c). The resulting image is binarized, and the intersection of this image with the original binary mask is found. This procedure takes advantage of the fact that the middle piece and tail are thinner than the spermatozoon head and therefore are blurred after Gaussian filtering. In Fig. 2a, b we show two examples of spermatozoa as they appear in the original stained sperm image and in Fig. 2c, d the corresponding distance image.

2.4 Spermatozoa Abnormality Classification

After segmentation, the sperm heads' morphological features are used for abnormality classification. We considered most of the shape descriptors proposed in the literature, including form factor (FF), roundness (R), solidity (S), extents1 (E1, defined as the ratio between the net area and the minimum bounding rectangle), extents2 (E2, defined as the ratio between net area and a circle area), eccentricity (ECC), aspect ratio (AR), convexity (C), and compactness (CC) [49]. These descriptors together were not able to discriminate properly among normal and abnormal sperm head shapes. For this reason we also investigated new bilateral and radial symmetry shape descriptors, including area and perimeter with respect to the main axis (SA_m and SP_m), area and perimeter with respect to the secondary symmetry axis (SA_s and SP_s), mean and standard deviation of areas (μ_A and σ_A), maximum and minimum areas (Max_A and Min_A), mean and standard deviation of perimeters (μ_P and σ_P), and maximum and minimum perimeters with respect to symmetry axes (Max_P and Min_P) [50]. These shape descriptors were used to train a SVM classifier [51]. Considering the advantages of applying the linear and radial basis mentioned in [52], we implemented two SVM classifiers: the first one (SVML) with linear kernel and the second one (SVMR) with radial basis kernel. The radial basis kernel maps the nonlinearly separable data examples to a higher-dimensional space in which the training data become separable. The linear kernel can handle the classification problem better when the relationship between tags and attributes is highly nonlinear. In summary, the complete workflow processes the sperm image, detects and segments each of the spermatozoa in the frame, extracts their morphological features, and classifies them as either normal or abnormal according to the classification model. The classification paradigms were elicited in close interaction with experts in male reproductive health. The training examples are both publicly available repositories and a set of sperm images privately produced in a dedicated health center. The sperm images in all the sample set were cropped and presented to three independent and highly trained specialists under different rotations to avoid geometrical bias. The specialists classified each sample either as normal or abnormal, in the latter case stating the reasons for the perceived abnormality (Fig. 3).

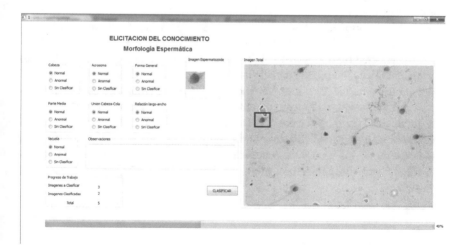

Fig. 3 User interface for sperm classification (the software interface is in Spanish for use of Spanish-speaking specialists)

2.5 Segmentation and Classification Performance Analysis

In segmentation, each pixel of the image was classified into two classes: background and sperm head. First, to evaluate the segmentation performance quantitatively, we used a confusion matrix [53, 54]. The parameters true positive (TP) (number of pixels in spermatozoa detected as such), false positive (FP) (number of non-spermatozoa pixels wrongly detected as pertaining to spermatozoa), false negative (FN) (number of spermatozoa pixels that were not detected), and true negative (TN) (number of non-spermatozoa pixels detected as not pertaining to spermatozoa) were computed. Since the sperm heads occupy a small proportion of the pixels in an image, then FN is usually much higher than the rest of the parameters in the confusion matrix. From these parameters, precision, true positive rate, and F-score are calculated, being F-score in general the most used in the evaluation of classifiers in this unbalanced class cases.

On the other hand, in the context of medical imaging, where there is seldom a ground truth against which an automated segmentation algorithm can be tested, the quality assessment is related to consensus with experts' opinion. In our case, the quality of the cross-out segmentation can be evaluated by measuring the spatial coincidence or overlaps of the automatically and manually segmented sperm. As ground-truth information we used the segmented plates provided by the gold standard dataset, and we computed the Dice similarity coefficient (DSC) [55] which is based on the ratio of the intersection of the two segmentation (automated and experts') with respect to the union. With respect to the ability of the model to assess as to whether a given spermatozoon is normal or abnormal, a quantitative evaluation was performed.

The aforementioned dataset containing 3196 sperm head samples, each with their corresponding 21 shape descriptors, was used to train a SVM automated classifier. We selected randomly 70% of the samples to train and the remaining 30% to test the classifier. The parameters C (free parameter of SVM) and γ (parameter of the radial basis kernel function) were adjusted using an automatic adjusting function, which finds within an exploration range the optimal parameter values in each run. In our case, the ranges were C in [1.0.20] and γ in [0.1;2.0]. We performed the experiment using a radial (SVRM) basis kernel functions on balanced classes. A stratified tenfold was used to evaluate the classifier performance. We used a confusion matrix and the parameters precision, true positive rate, and F-score were estimated.

3 Results and Discussion

3.1 Segmentation

An intermediate result of the algorithm is a distance image, in which the distance in YIQ color space of every pixel in the original image with respect to the background color prototype of the corresponding stain is represented in a grayscale pseudocolor.

On this image, the darkest pixels correspond to the most similar in color to the background stain in the original sperm image, and by contraposition, the brighter ones are likely to be part of the sperm heads. Instead of this *a contrario* segmentation (based on background prototypes), a foreground-based segmentation might be performed, but the results are very unstable given the inherent color drift in the stain of the sperm heads and the comparatively small amount of pixels corresponding to sperm heads in a sperm image. In other words, the foreground prototype is too scarce and variable to be robust. Then the distance image histogram is computed and a brightness (distance) threshold is found using the maximum likelihood criterion. This criterion produces a few false negatives (sperm head pixels classified as background) and some false positives (non-sperm head pixels classified as such). The former are almost always due to the digitization procedure in the imaging, are located evenly around the spermatozoon head border, and therefore do not exert any significant influence in the sperm head morphology assessment. The false positives, instead, are mostly located in the middle piece of the spermatozoon, given that the stain therein is similar to the stain in the sperm head. For this reason, the mask procedure explained in the previous section must be applied. Figure 4 shows the results of applying our method to three images of the dataset of Shaker et al. [27]. Figure 5 shows a comparison of automatic and manual head segmentation. The segmentation procedure achieved a precision of 95%, a true positive rate of 95%, and a F-score of 96%, and the DC was 0.96.These results are similar to the ones reported by Shaker

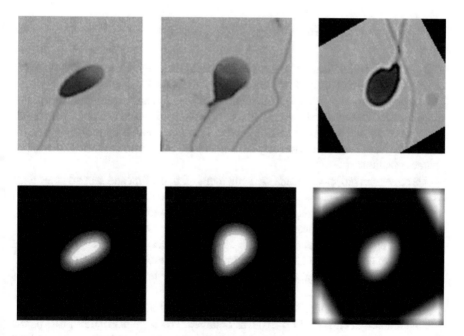

Fig. 4 Segmentation stage. (**a**) Original sperm image. (**b**) Gaussian blur

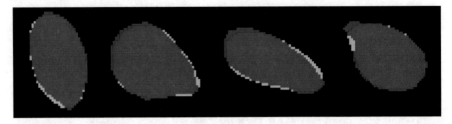

Fig. 5 Comparison of automatic and manual head segmentation. Red pixels show the overlap among both segmentation. In green are shown the pixels belonging to ground truth and in blue the ones segmented with the automatic method

Fig. 6 Three segmented sperm head samples, (**a, b**) normal and (**c**) abnormal, with the main and secondary symmetry axes superimposed in red and green, respectively

et al. [27] and by Chang et al. [26], which are considered the state of the art in automatic sperm head segmentation.

3.2 Sperm Head Morphology Classification

This part of the classification procedure is focused on analyzing the morphology of the segmented sperm heads. Over this representation, the 21 shape descriptors mentioned above are applied. Given that the achievable spatial resolution of sperm image renders sperm heads with a coarse digitalization, most conventional shape descriptors are not sufficient in differentiating normal and abnormal sperm head shapes. Instead, our novel symmetry descriptors are successful in sperm head shape characterization.

Three segmented sperm heads are shown in Fig. 6, the first and second identified as normal by the specialists and the third as abnormal. In Table 1 the results of applying the nine most frequently used shape descriptors to these examples are shown. The information provided by these descriptors is not sufficient to train an automated classifier that performs with adequate accuracy and precision. Instead, the results of applying the 12 novel symmetry shape descriptors, shown in Table 2, together with the traditional ones (21 shape descriptors in all), yield a more significant discriminative power.

In the classification stage, considering a tenfold validation an accuracy of 92% with a standard deviation of ($^+_-$0.10) was obtained. Precision was 91.5%, true positive rate was 92.5%, and F1-score was of 92%. The whole automatic processing

Table 1 Shape descriptors most frequently used in the literature, applied to the examples in Fig. 6

Sample	FF	R	AR	E1	S	E2	ECC	C	CC
(a)	0.621	0.743	0.685	0.815	0.933	0.748	0.738	0.861	0.741
(b)	0.635	0.745	0.726	0.769	0.937	0.745	0.687	0.858	0.747
(c)	0.578	0.701	0.711	0.739	0.914	0.701	0.702	0.832	0.725

Table 2 Symmetry shape descriptors applied to the examples in Fig. 6

Sample	SA_m	SP_m	μ_A	σ_A	Max_A	Min_A	μ_P	σ_P	Max_P	Min_P	SA_S	SP_S
(a)	0.998	0.992	0.838	0.106	0.983	0.75	0.912	0.078	0.972	0.838	0.97	0.93
(b)	0.985	0.996	0.936	0.036	0.992	0.9	0.94	0.043	0.993	0.855	0.92	0.97
(c)	0.919	0.978	0.848	0.069	0.927	0.743	0.961	0.051	0.992	0.992	0.96	0.97

(segmentation and classification) achieved an average precision of 93.25%, average TPR of 93.75%, and average F-score of 94%.The AUC of the classifier is 0.92 regarding normal/abnormal sperm classification.

4   Conclusion and Future Work

The main contributions of this work are (i) an approach to sperm head segmentation that preserves their shape without losing the original morphology even with strong stain variations and unevenness produced during image acquisition, (ii) a novel descriptor of sperm head morphology, and (iii) a SVM-based classification model for sperm head morphology. The final framework is fully autonomous and therefore may be deployed as a web service that provides automatic determination of sperm head morphology in humans based upon samples received over the Internet.

The image processing framework is able to segment sperm heads from regular stained microscopy images and classify them as normal or abnormal according to their morphology. A pre-segmentation step is performed matching the color histogram of the sample to color histograms of known stains. This color-based matching of foreground and background color prototypes performs a pixel-wise detection of the sperm heads present in the sample. This pre-segmentation is not completely accurate, since pixels corresponding to the middle piece sometimes get stained similarly to the sperm heads.

For this reason, a post-segmentation procedure is applied using first a Gaussian filter and then a binarization mask. Results show an improvement of 4% in pixel-wise segmentation with respect to human-assisted segmentation (ground truth), which appears to be above other results reported in the literature. An additional advantage of this procedure is that it preserves the edges and overall shape of the spermatozoa as they are perceived by expert humans.

After segmentation, a classification procedure is applied distinguishing two classes (normal or abnormal) based on a set of 21 traditional and novel shape descriptors. Each sperm head is then characterized with a feature vector with these 21 values, which describe relevant morphological aspects in terms of their normality. A SVM classifier was trained using the public HuSHeM dataset and the gold standard dataset. The whole automatic processing (segmentation and classification) achieved an average precision of 93.25%, average TPR of 93.75%, and average F-score of 94%.

The AUC of the classifier is 0.92 regarding normal/abnormal sperm classification.

We are currently working on extending these results in several lines. First, a finer analysis of different abnormal conditions can be performed applying more shape descriptors to the sperm head. Second, we are developing sperm acrosome and tail segmentation and description, leading to a much richer understanding of the overall sperm condition. Third, we plan to integrate the shape descriptors of spermatozoa in the sample image as a feature to evaluate semen analysis. Also, we are applying other machine learning techniques (for instance deep learning) as an alternative to SVM. Finally, a web-based fully automated CASA system is currently under development. This can be used both for scientific and clinical purposes, and the information accrual of a large number of samples may trigger fruitful demographic investigations.

Acknowledgments The authors thank VITA Medicina Reproductiva (IPAMER SRL) laboratory, Chubut, Argentina. The project was partially funded by Secretaría General de Ciencia y Técnica, Universidad Nacional del Sur, Argentina (grants PGI 24/K083, PGI 24/ZK26), and National Council of Science and Technology of Argentina.

References

1. WHO laboratory manual for the examination and processing of human semen, World Health Organization, 5th edn. (2010)
2. E. Levitas, E. Lunenfeld, N. Weisz, M. Friger, G. Potashnik, Relationship between age and semen parameters in men with normal sperm concentration: analysis of 6022 semen samples. Andrologia **39**(2), 45–50 (2007)
3. K.P. Nallella, R.K. Sharma, N. Aziz, A. Agarwal, Significance of sperm characteristics in the evaluation of male infertility. Fertil. Steril. **85**(3), 629–634 (2006)
4. W. Ombelet, R. Menkveld, T.F. Kruger, O. Steeno, Sperm morphology assessment: historical review in relation to fertility. Hum. Reprod. Update **1**(6), 543–557 (1995)
5. M. Hidalgo, I. Rodríguez, J. Dorado, Influence of staining and sampling procedures on goat sperm morphometry using the sperm class analyzer. Theriogenology **66**(4), 996–1003 (2006)
6. M.L. Poland, K.S. Moghissi, P.T. Giblin, J.W. Ager, J.M. Olson, Variation of semen measures within normal men. Fertil. Steril. **44**(3), 396–400 (1985)
7. N.G. Berman, C. Wang, C.A.L.V.I.N. Paulsen, Methodological issues in the analysis of human sperm concentration data. J. Androl. **17**(1), 68–73 (1996)
8. E. Carlsen, J.H. Petersen, A.-M. Andersson, N.E. Skakkebaek, Effects of ejaculatory frequency and season on variations in semen quality. Fertil. Steril. **82**(2), 358–366 (2004)
9. J.L. Yániz, S. Vicente-Fiel, S. Capistrós, I. Palacín, P. Santolaria, Automatic evaluation of ram sperm morphometry. Theriogenology **77**(7), 1343–1350 (2012)

10. M. Ramón, F. Martínez-Pastor, O. García-Alvarez, A. Maroto-Morales, A.J. Soler, P. Jiménez-Rabadán, M.R.F. Santos, R. Bernabéu, J.J. Garde, Taking advantage of the use of supervised learning methods for characterization of sperm population structure related with freezability in the Iberian red deer. Theriogenology **77**(8), 1661–1672 (2012)

11. C. Soler, T.G. Cooper, A. Valverde, J.L. Yániz, Afterword to sperm morphometrics today and tomorrow special issue in Asian Journal of Andrology. Asian J. Androl. **18**(6), 895 (2016)

12. J. Verstegen, M. Iguer-Ouada, K. Onclin, Computer assisted semen analyzers in andrology research and veterinary practice. Theriogenology **57**(1), 149–179 (2002)

13. M. Adamkovicova, R. Toman, M. Cabaj, S. Hluchy, P. Massanyi, N. Lukac, M. Martiniaková, Computer assisted semen analysis of epididymal spermatozoa after an interperitoneal administration of diazinon and cadmium. Sci. Papers Animal Sci. Biotechnol. **45**(1), 105–110 (2012)

14. R. Menkveld, F.S.H. Stander, T.J.W. Kotze, T.F. Kruger, J.A. van Zyl, The evaluation of morphological characteristics of human spermatozoa according to stricter criteria. Hum. Reprod. **5**(5), 586–592 (1990)

15. M.J. Tomlinson, E. Kessopoulou, C.L.R. Barratt, The diagnostic and prognostic value of traditional semen parameters. J. Androl. **20**(5), 588–593 (1999)

16. F.H. Van der Merwe, T.F. Kruger, S.C. Oehninger, C.J. Lombard, The use of semen parameters to identify the subfertile male in the general population. Gynecol. Obstet. Investig. **59**(2), 86–91 (2005)

17. A. Cipak, P. Stanić, K. Đurić, T. Serdar, E. Suchanek, Sperm morphology assessment according to who and strict criteria: method comparison and intra-laboratory variability. Biochemia Medica **19**(1), 87–94 (2009)

18. J.F. Moruzzi, A.J. Wyrobek, B.H. Mayall, B.L. Gledhill, Quantification and classification of human sperm morphology by computer assisted image analysis. Fertil. Steril. **50**(1), 142–152 (1988)

19. R.O. Davis, D.E. Bain, R.J. Siemers, D.M. Thal, J.B. Andrew, C.G. Gravance, Accuracy and precision of the cellform-human automated sperm morphometry instrument. Fertil. Steril. **58**(4), 763–769 (1992)

20. K. Coetzee, T.F. Kruger, C.J. Lombard, Repeatability and variance analysis on multiple computer-assisted (IVOS*) sperm morphology readings. Andrologia **31**(3), 163–168 (1999)

21. K.S. Park, W.J. Yi, J.S. Paick, Segmentation of sperms using the strategic Hough transform. Ann. Biomed. Eng. **25**(2), 294–302 (1997)

22. W.J. Yi, K.S. Park, J.S. Paick, Parameterized characterization of elliptic sperm heads using Fourier representation and wavelet transform. In *Engineering in Medicine and Biology Society, 1998. Proceedings of the 20th Annual International Conference of the IEEE*, vol. 2, pp. 974–977, IEEE (1998)

23. H. Carrillo, J. Villarreal, M. Sotaquira, A. Goelkel, R. Gutierrez, A computer aided tool for the assessment of human sperm morphology. In *Bioinformatics and Bioengineering, 2007. BIBE 2007. Proceedings of the 7th IEEE International Conference on*, pp. 1152–1157, IEEE (2007)

24. V.S. Abbiramy, V. Shanthi, Spermatozoa segmentation and morphological parameter analysis based detection of teratozoospermia. Int. J. Comput. Appl. **3**(7), 19–23 (2010)

25. A. Bijar, A. Pe, M. Mikaeili, Fully automatic identification and discrimination of sperm's parts in microscopic images of stained human semen smear. J. Biomed. Sci. Eng. **05**(07), 384–395 (2012)

26. V. Chang, J.M. Saavedra, V. Castañeda, L. Sarabia, N. Hitschfeld, S. Hartel, Gold-standard and improved framework for sperm head segmentation. Comput. Methods Prog. Biomed **117**(2), 225–237 (2014)

27. F. Shaker, S.A. Monadjemi, A.R. Naghsh-Nilchi, Automatic detection and segmentation of sperm head, acrosome and nucleus in microscopic images of human semen smears. Comput. Methods Prog. Biomed. **132**, 11–20 (2016)

28. V. Chang, L. Heutte, C. Petitjean, S. Härtel, N. Hitschfeld, Automatic classification of human sperm head morphology. Comput. Biol. Med. **84**, 205–216 (2017)

29. M. Ramón, F. Martínez-Pastor, Implementation of novel statistical procedures and other advanced approaches to improve analysis of casa data. Reprod. Fertil. Dev. (2018)
30. F. Ghasemian, S.A. Mirroshandel, S. Monji-Azad, M. Azarnia, Z. Zahiri, An efficient method for automatic morphological abnormality detection from human sperm images. Comput. Methods Prog. Biomed. **122**(3), 409–420 (2015)
31. K.-K. Tseng, Y. Li, C.-Y. Hsu, H.-N. Huang, M. Zhao, M. Ding, Computer-assisted system with multiple feature fused support vector machine for sperm morphology diagnosis. Biomed. Res. Int. **2013**, 687607 (2013)
32. J.L. Yániz, S. Capistrós, S. Vicente-Fiel, C. Soler, J.N. de Murga, P. Santolaria, Study of nuclear and acrosomal sperm morphometry in ram using a computer-assisted sperm morphometry analysis fluorescence (CASMA-F) method. Theriogenology **82**(6), 921–924 (2014)
33. P. Santolaria, C. Soler, P. Recreo, T. Carretero, A. Bono, J.M. Berné, J.L. Yániz, Morphometric and kinematic sperm subpopulations in split ejaculates of normozoospermic men. Asian J. Androl **18**(6), 831 (2016)
34. F. Vásquez, C. Soler, P. Camps, A. Valverde, A. García-Molina, Spermiogram and sperm head morphometry assessed by multivariate cluster analysis results during adolescence (12–18 years) and the effect of varicocele. Asian J. Androl. **18**(6), 824 (2016)
35. S. Sadeghi, A. García-Molina, F. Celma, A. Valverde, S. Fereidounfar, C. Soler, Morphometric comparison by the ISAS® CASADNAF system of two techniques for the evaluation of DNA fragmentation in human spermatozoa. Asian J. Androl. **18**(6), 835 (2016)
36. M.A. Gutiérrez-Reinoso, M. García-Herreros, Normozoospermic versus teratozoospermic domestic cats: differential testicular volume, sperm morphometry, and subpopulation structure during epididymal maturation. Asian J. Androl. **18**(6), 871 (2016)
37. H. Cucho, V. Alarcón, C. Ordóñez, E. Ampuero, A. Meza, C. Soler, Puma (puma concolor) epididymal sperm morphometry. Asian J. Androl. **18**(6), 879 (2016)
38. M. García-Herreros, Sperm subpopulations in avian species: a comparative study between the rooster (gallus domesticus) and Guinea fowl (numida meleagris). Asian J. Androl. **18**(6), 889 (2016)
39. SCA, http://www.micropticsl.com/es/productos/sca-sistema-casa/, last accessed 2020/10/15
40. ISAS, http://www.proiser.com/en/}, last accessed 2020/10/15
41. IVOS   http://www.hamiltonthorne.com/index.php/products/clinical-casa-products/ivos-ii-clinical, last accessed 2020/10/15
42. S. Gunalp, C. Onculoglu, T. Gurgan, T.F. Kruger, C.J. Lombard, A study of semen parameters with emphasis on sperm morphology in a fertile population: an attempt to develop clinical thresholds. Hum. Reprod. **16**(1), 110–114 (2001)
43. T.B. Haugen, T. Egeland, Ø. Magnus, Semen parameters in Norwegian fertile men. J. Androl. **27**(1), 66–71 (2006)
44. L. Fraser, C.L. Barratt, D. Canale, T. Cooper, C. DeJonge, S. Irvine, D. Mortimer, S. Oehninger, J. Tesarik, Consensus workshop on advanced diagnostic andrology techniques. Eshre Andrology Special Interest Group. Hum. Reprod. **12**(4), 873–873 (1997)
45. M.J. Zinaman, M.L. Uhler, E. Vertuno, S.G. Fisher, E.D. Clegg, Evaluation of computer-assisted semen analysis (CASA) with ident stain to determine sperm concentration. J. Androl. **17**(3), 288–292 (1996)
46. C. Garrett, D.Y. Liu, G.N. Clarke, D.D. Rushford, H.W.G. Baker, Automated semen analysis: zona pellucida preferred sperm morphometry and straight-line velocity are related to pregnancy rate in subfertile couples. Hum. Reprod. **18**(8), 1643–1649 (2003)
47. N.V. Revollo, C.A. Delrieux, G.M.E. Perillo, Automatic methodology for mapping of coastal zones in video sequences. Mar. Geol. **381**, 87–101 (2016)
48. Girard, Michel-Claude and Girard, Colette and Courault, Dominique and Gilliot, Jean-Marc and Loubersac, Lionel and Meyer-Roux, Jean and Monget, Jean-Marie and Seguin, Bernard and Rao, N Venkat, Processing of remote sensing data, Routledge (2010).
49. Neal, F. Brent, and John C. Russ. Measuring shape. CRC Press (2012)

50. N.V. Revollo, C.A. Delrieux, R. González-José, Set of bilateral and radial symmetry shape descriptor based on contour information. IET Comput. Vis. **11**(3), 226–236 (2016)
51. Vapnik, Vladimir. The nature of statistical learning theory. Springer Science & Business Media (2013)
52. C.-W. Hsu, C.-C. Chang, C.-J. Lin, et al., A practical guide to support vector classification, (2003)
53. M. Story, R.G. Congalton, Accuracy assessment: a user's perspective. Photogramm. Eng. Remote. Sens. **52**(3), 397–399 (1986)
54. G.M. Foody, Status of land cover classification accuracy assessment. Remote Sens. Environ. **80**(1), 185–201 (2002)
55. L.R. Dice, Measures of the amount of ecologic association between species. Ecology **26**(3), 297–302 (1945)

Future Contribution of Artificial Vision in Methodologies for the Development of Applications That Allow for Identifying Optimal Harvest Times of Medicinal Cannabis Inflorescences in Colombia

Luis Octavio González-Salcedo ⓘ, Andrés Palomino-Tovar, and Adriana Martínez-Arias ⓘ

1 Introduction

Artificial or computational vision has presented great challenges in this twenty-first century, and its applications include different areas where agriculture, medicine, domestic and defense, marine and space exploration, manufacturing, and transport, among others, stand out [1]. New technologies require intelligent machines that are capable of allowing new applications for different domains and services. Different authors have advanced in terms of research allowing these technologies to be available to the populations through software that can be executed on machinery, computers, and even mobile devices.

Advances in visual attention models increasingly bring technology closer to being able to differentiate more defined details in images, as established in the "content-based image retrieval model" proposed by [2], where they attempt to more precisely simulate the mechanisms of visual attention that help humans and primates quickly select the most relevant information from a scene. This model develops three highlights: (1) include color volume with edge information (identification of points that have a drastic change in brightness of a digital image) as a new visual cue to detect prominent regions in the study object; (2) use gray-level matching matrices, to suppress global maps using texture recognition through the differences between energy characteristics that objects in an image can produce; and (3) propose an image representation method called a prominent structure histogram where the selective orientation mechanism for image representation is stimulated.

On the other hand, image processing through artificial vision is helping, among many occupations, the health professional in the analysis of biomedical images to

L. O. González-Salcedo (✉) · A. Palomino-Tovar · A. Martínez-Arias
Facultad de Ingeniería y Administración, Universidad Nacional de Colombia Sede Palmira, Palmira, Colombia
e-mail: logonzalezsa@unal.edu.co

© Springer Nature Switzerland AG 2022
P. Johri et al. (eds.), *Trends and Advancements of Image Processing and its Applications*, EAI/Springer Innovations in Communication and Computing, https://doi.org/10.1007/978-3-030-75945-2_10

obtain more reliable diagnoses by implementing vision algorithms in the development of applications, for focused software in the visualization and processing using a visualization toolkit, which corresponds to a set of freely distributed code libraries that have object-oriented programming [3]. This tool can be used to present software capable of 3D reconstruction from a stack of 2D biomedical images.

Reference [4] explain how high-resolution video cameras are used to help the BRAiVE vehicle to transport itself autonomously through image processing; lane detection is performed in six steps. First, a transformation from inverse perspective mapping to the camera image capture is performed. Second, a low-level filtering that compares light-dark patterns is carried out. Third, the resulting points are grouped and these groups are approximated by continuous functions for line segments. Fourth, the produced line segment lists against existing lane markings from a previous stage of tracking are compare. Fifth, several brand candidates that are compared with the acceptance thresholds are generated. Sixth, the vehicle decides which lane markings to follow.

In relation to image recognition and artificial intelligence, [5] show the training of an artificial neural network to diagnose the stress intensity produced by excess water in tomato plants. Training is done by identifying some different morphological characteristics that the flower adopts when under these stress conditions, including the increased number of petals, the oval shape of the anther cone, and the appearance of two stigmas instead of one. A multilayer neural network for image recognition was used, and their training for learning was supervised by applying the backward propagation calculation method or backpropagation algorithm.

Reference [6] generates a computer vision system, under the "incremental cascade" software development model, which allows quantifying and identifying four pathogenic bacteria present in food in a culture medium, through pattern recognition. related to the differences between their morphological characteristics. This research used a neural network trained under supervised learning, with a database made with processed images, including noise elimination, grayscale filtering, enhancement by linear expansion of the histogram to enlarge the gray range and its resolution, and segmentation to increase the contrast with the image background. The aforementioned procedure allowed the counting of regions where the image is inverted and colonies are labeled, RGB segmentation to eliminate the culture medium and imaging only bacterial colonies, obtaining color to identify bacteria according to the database, and finally the simulation to check the operation of the application.

Considering that the fields of computational vision, artificial vision, and artificial intelligence are used successfully for pattern recognition, it is desired to promote technological development for a crop of significant economic and scientific interest in Colombia: medicinal cannabis. The little research that there is in the country about this plant makes it necessary to direct efforts toward the studies of its different phenological stages and highlight the importance of certain practices to generate high-quality products, such as determining the optimum moment of maturity to harvest. In cannabis harvest, the proportion of cannabinoids varies according to an early or late harvest, in such a way that a precise identification of the harvest time is necessary to standardize the quality of the product and avoid losses of the compounds to be extracted. Other problems are identified in the late harvest, for example, tetra-hydro-cannabinoids and cannabidiols are degraded into less interesting compounds such as cannabinols.

This research explores the necessary components for the future development of an application designed for the recognition of the optimal harvest time of *Cannabis* sp. inflorescences, using technologies based on artificial intelligence, for the recognition of images that classify and count the maturation states in trichomes.

1.1　Inflorescence of **Cannabis** *sp. Maturation and Optimal Harvest Time*

Generalities About the Inflorescences of *Cannabis* sp.

Because *Cannabis* sp. is a dioecious plant, its female and male inflorescences occur in different individuals and there are some exceptions that reveal hermaphroditism in some specimens [7], and a general taxonomic approach is required in both types of inflorescences. Figure 1 shows the generalities of the *Cannabis* sp. inflorescences, both in male inflorescences and in female inflorescences.

Due to the dioecious nature of *Cannabis* sp., there are two types of inflorescences, female and male: the first, which contain the ovules connected to the pistils, and the second, which provide pollen through their stamens. Initially visual identification is not an easy task, but with the growth of the inflorescences, the changes become more evident [9]. The male and female inflorescences of *Cannabis* sp. plants are not different only because of their reproductive organ but also because of their differences in maximum heights, since the former tend to be between 10% and 15% higher than the latter, and in addition to this, they tend to be less robust with thinner and less branched stems [9]. As the interest of visual identification is focused on female inflorescences, Table 1 makes a general approach to the processes that occur in the week-by-week development of the flowering of some varieties. It should be noted that depending on the variety, nutrition, and climate, among other factors, times may change [10].

Plant Structures of Interest

Pistils

The pistils are female organs of the flowers and inflorescences that have the function of transporting pollen toward the ovary once it is captured by the stamen; they are generally shaped like "hairs," and in the female inflorescences of *Cannabis* sp., they protrude from the calyx and can reach lengths of up to 3 mm [11]. The pistils are present in the different development phases of the *Cannabis* sp. plant, from the

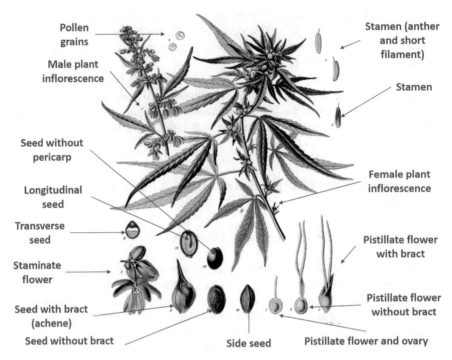

Fig. 1 Generalities about *Cannabis* sp. inflorescences. (Adapted from Ref. [8], public domain image available at https://commons.wikimedia.org/wiki/File:Cannabis_sativa_-_K%C3%B6hler%E2%80%93s_Medizinal-Pflanzen-026.jpg)

Table 1 Week-by-week development of the flowering stage of *Cannabis* sp.

Week	Description
1	Stretching for the production of structure for the inflorescences
2	The first shoots of white pistils appear
3	Formation of small buds, growing slower and slower
4	Plant growth stops; pistils reach larger sizes
5	Bud thickening; significant increase in odor and pistils begin to darken and start the expected coloration
6–8	Final stage of flowering; orange, pink, or red pistils, among other color; maximum thickening of the bud

Own Elaboration, adapted from Ref. [10]

vegetative phase through the pre-flowering, flowering, and ripening phases; in the latter, its tones transform from white to orange and red or pink due to physical-chemical changes [7].

Trichomes

Trichomes are present in a large number of plants and have different functions: absorb water, protect against different solar rays, regulate temperature, repel and attract insects, and generate metabolites of great importance such as cannabinoids, flavonoids, and terpenes [12]. Sessile glandular trichomes have a globose head measuring between 30 and 50 μm; are found mainly on the surfaces of petioles, the underside of leaves, and young stems; and are most visible in the early stages of bract development [13]. Bulbous trichomes are smaller and are present in *Cannabis* sp. plants, they measure between 10 and 20 μm and their height exceeds 30 μm, and they have a small head a little wider than their stem [13]. The burst or capitate glandular trichomes have a globose head like the sessiles that reach larger sizes between 50 and 70 μm and additionally have a fairly robust stem that can be between 100 and 200 μm [14].

Other types of trichomes are also present: non-glandular trichomes. Simple unicellular trichomes structurally similar to a "hair" are produced initially in the cotyledons and continue to develop on the underside of the leaves, stems, and petioles, and they can measure between 250 and 370 μm [15]. Systolic trichomes are found on the surface of the leaves and have a rough texture, they are shorter than simple unicellular trichomes, their dimensions are between 10 and 125 μm, and additionally they have a base that oscillates between 60 and 140 μm; they have functions as insecticides since they produce substances that affect ingestion in pests, and these trichomes do not have cannabinoids [7].

Harvest, Maturation, and Harvest Time for *Cannabis* sp. Inflorescences

The maturation stage is a process involving multiple changes at the cellular level reflected in different characteristics in addition to the increase in size [16]; this stage requires the synthesis of proteins and messenger RNA and these processes require a lot of energy to develop, which is obtained from different sets of biochemical reactions that give rise to cellular respiration [17]. In this study, the distinction that exists in idiomatic terms between maturity and harvest time is highlighted: the former is aimed at those fruits where the optimum harvest point is reached when physical and biochemical changes in shape, size, and texture can be detected. Color, smell and taste [18]. The second is applied when the harvested part of the plant is not the fruit [19] or for fruits that do not manage to undergo the aforementioned transformations [18]. Since in the cultivation of cannabis for medicinal and scientific purposes the inflorescence is collected, it is understood that the ideal term to speak in cultivation is that of the moment of harvest, and for explanatory purposes, it refers to maturity at the moment of harvest.

For the crop of interest, it is essential to note that because the time of harvest occurs before the development of the inflorescence, it must survive after being harvested with its own collected substrates, that is, the transition that exists between the completion of growth and the beginning of senescence [20]. At the time of harvest, the inflorescences of *Cannabis* sp. suffer from changes in their coloration due to the modification of chlorophyll and carotenoid contents that generate an accumulation of flavonoids; likewise, the textural transformation occurs due to the incidence of the alteration of the cell wall due to turgor and the metabolism of fatty acids, sugars, and volatile compounds, and this change can improve or harm the taste, aroma, or nutritional compounds of the final product [21]. Some varieties of *Cannabis* sp. have a uniform ripening process for all their inflorescences, which makes the identification and the harvest process easier; but there are varieties that have a discontinuous maturation, from top to bottom, from the inside out; however, generally the inflorescences most favored by light ripen first and these, once harvested, give way to the maturation of the other inflorescences [7].

Types of Maturity

For the cultivation of *Cannabis* sp., there are two types of maturity: physiological maturity, which refers to the end of physiological development, and commercial maturity, which includes the necessary steps to have the product ready in its biological state either for consumption or transformation [18]. Physiological maturity is the stage where the development of the plant material of interest reaches its maximum growth and maturation; in this phase the inflorescences of *Cannabis* sp. reach their physical and biological development according to their natural cycle and, after this moment, the senescence stage [18]. In the process of maturation of *Cannabis* sp. inflorescences, different physicochemical changes occur where chemical reactions occur that alter the contents of the different compounds that compose them.

Visual and Touch Indicators of Maturation States

Although the inflorescences of *Cannabis* sp. undergo transformations due to the biochemical processes mentioned above, the visual identification of some of these changes is not as evident as in most fruits, but the presence of certain signs that may be perceptible to the human eye and touch and observable with the help of some instruments [7]. Table 2 shows the observable visual characteristics and a brief description of them.

The most important indicator to identify the optimal harvest time in inflorescences of *Cannabis* sp. is associated with trichomes; these are mainly found in the inflorescences and around the leaves, and they are also present in stems and leaves [15]. The types of trichomes to be evaluated to define the harvest time are the burst or capitate glandular trichomes, since these contain higher concentrations of cannabinoids. Initially, these appear translucent or transparent due to their lack of chloroplasts, but as they mature, they turn opaque white until they reach an amber or brown color that indicates the degradation of some of the cannabinoids, that is, that are overmature [12].

Table 2 Visual and tactile indicators to identify the time of harvest

Characteristic	Description
Yellowing of the leaves (senescence)	Senescence is the irreversible genetic programming that leads to the death of a plant, leaf, tissue, or cell, the final stage of development [11]
Leaf fall (abscission)	The abscission is the natural fall of the leaves that give way to the formation of new structures [11]
Stiffness (turgor)	The high moisture content exerts pressure on the plant cell walls, which creates rigidity to the plant; at the time of harvesting the *Cannabis* sp., the moisture content is around 80% [7]
Stem desiccation	The stem conducts water and nutrients from the roots to the leaves and flowers; in some species the inflorescences gradually absorb the nutrients from the stem generating a progressive desiccation [7]
Resin production	The inflorescences of *Cannabis* sp. are covered with resin glands, generating a sticky sensation to the touch [7]
Swollen chalice	Near maturity, the inflorescences of *Cannabis* sp. show a swelling in the caps of their calyx due to the accumulation of resin [7]
Pistils with red and pink hues around 70–75%	When the inflorescence of *Cannabis* sp. is at its maximum point of maturation, the pistils turn from white to orange and red or pink, and when they reach a proportion between 70% and 75%, it can be harvested [11]

Own Elaboration, adapted from Refs. [7, 11]

1.2 Artificial Intelligence Applied to Image Recognition

Artificial intelligence is the discipline that tries to emulate some of the intellectual capacities of the human brain and the capacities of living beings to make decisions and solve various problems, using artificial systems supported by computational theories. Sensory perception processes such as vision and hearing, among others, and their consequent pattern recognition processes are what are commonly known as human intelligence. For this reason, the most common applications of artificial intelligence are process data and identify systems [22]. For the solution of a problem, an artificial system needs a finite sequence of instructions that specify the different actions to be executed by the computational medium, and this sequence of instructions forms the algorithmic structure of the system, in charge of reducing the problem to a set of rules [22].

One of the areas where advances have been most notable is image recognition, partly thanks to the development of new deep learning techniques. However, various artificial intelligence techniques are used in digital image processing, including fuzzy systems, artificial neural networks, swarm intelligence, and artificial immune systems. Both for the conventionally used techniques and for the new ones toward which deep learning tends, [23] define that image processing addresses three consecutive stages: thresholding, cleaning, and filtering. Each of these stages develops a series of operations to obtain the required information using RGB photographs, that is, three-color multilayer photographs. In these operations, thresholding selects intensity ranges and manipulates the histogram values of each layer, cleaning obtains defined shapes by applying different morphological operations to remove noise and unnecessary elements, while feature filtering performed through spatial analysis seeks to find objects and geometries [24, 25].

Zadeh presented in 1965 the theory of fuzzy sets [26] with which he founded fuzzy systems, which have a great application in engineering, mainly in the areas of digital image processing and control. In the specific field of image recognition, fuzzy sets are used because they do not limit the process numerically, being useful on images where they cannot be delimited in an exact way [27]. The brain is made up of biological neurons and specialized regions that store, process, and characterize a large amount of information, learning behavior that is emulated by artificial neural networks in a computational environment [28]. This neuronal simile allows the generation of complex systems by creating a structure where simple elements are used as information storage variables, connectivity through differentiable functions to be adjusted by error correction in their estimation, and system calibration processes using indicators such as the quantification of the error or regression fit in the results obtained [29]. Various types of artificial neural models have been developed, the most applied in image recognition being multilayer perceptron networks, Kohonen networks, deep belief networks, recurrent networks, and convolutional networks, the latter being the ones that in recent years have specialized in working with images [30, 31].

With regard to swarm intelligence, this corresponds to granting an artificial system property inspired by biological groups or nature, in which joint actions are carried out to fulfill a specific task [32]. Some models have been implemented based on the behavior and functioning of collective work groups such as colonies of ants and bees, bacterial growth, groups of birds and fish, grazing of animals, and the movement of particle clouds [33]. In optimization and image analysis tasks, bee colonies correspond to one of the most widely used swarm systems for the aforementioned purposes [34]. Finally, the immune system can be defined as the one in charge of protecting an organism from external elements that can attack it, before which it acts by recognizing and eliminating those attacking elements [35]. In this way, artificial systems try to emulate this behavior for problem solving using the concept of antibody production based on the information collected and saved [36, 37]. In the field of image recognition, these types of models have been used primarily in the interpretation of medical images for various anomalies and pathologies.

Multilayer Perceptron Artificial Neural Network (MLP ANN)

As mentioned in the previous paragraph, an artificial neural network is an information processing paradigm that is inspired by the way a biological nervous system processes information [38]. It is made up of a large number of highly interconnected processing elements (neurons) that work together to solve specific problems and can be used for classification or regression [38]. The multilayer perceptron network

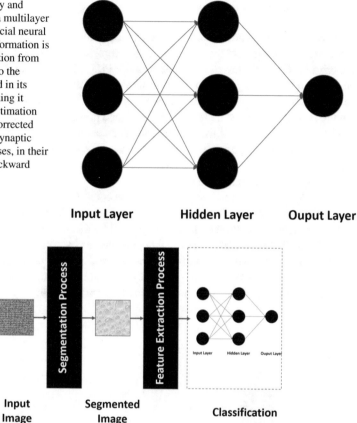

Fig. 2 Typology and architecture of a multilayer perceptron artificial neural network: the information is fed in one direction from the input layer to the output layer, and in its supervised learning it calculates the estimation error which is corrected (applied to the synaptic weights and biases, in their connections) backward

Fig. 3 Learning method of a multilayer perceptron applied in image recognition. (Own Elaboration)

is one of the simplest types of artificial neural networks in which the information moves in a single direction – forward – from the input nodes, and the nodes pass through the hidden nodes and toward the nodes of outlet [38], as shown in Fig. 2.

The training of a multilayer neural network is carried out through a learning process in which the approximation in the recognition of the patterns is made from a backward propagation learning algorithm known as the backpropagation algorithm, in which the connections and adjustments (called synaptic weights and biases) are adjusted during this process [39]. Multilayer perceptron is a supervised learning network, that is, it requires knowing the outputs that are included in the training set, to correct and distribute the estimation error backward [39]. In this type of multilayer network, the methodology for image recognition consists of two parts, a segmentation process and an extraction process [40], as shown in Fig. 3.

Convolutional Neural Networks

Convolutional neural networks are networks especially used in the field of artificial vision because their neurons correspond to receptive fields, such as the neurons of the primary visual cortex of a biological brain [41]. These networks are especially effective in classifying and segmenting images and any other type of data that is continuously distributed throughout the input layer. Convolution in broad strokes is the process of adding each element of the image powered by a nucleus to its local neighbors. The kernel fulfills certain tasks, such as detection of vertical edges and horizontal edges, among others. Then, the main objective of the components of these networks is to perform the convolution operation in order to extract characteristics from the input image [42].

Convolutional neural network architectures consist of multiple layers of convolutional filters of one or more dimensions, and in each layer a function is usually added to perform a nonlinear causal mapping [43, 44]. In its most common version, generally dedicated to classification, it begins with a characteristic extraction phase that is carried out from convolutional neurons and sample reduction. In the final phases, neurons in a multilayer perceptron network perform classification on the extracted features [44, 45]. This process can be seen in Fig. 4.

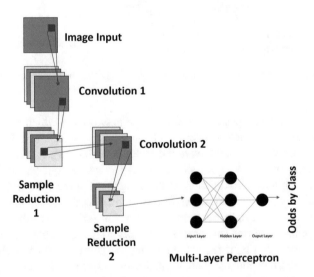

Fig. 4 Classical structure of a convolutional network for classification. (Own Elaboration)

2　Background on Image Recognition for Similar Applications

The development and training of artificial neural networks for image recognition is one of the applications in the field of artificial intelligence. The literature reports that in the works related to image processing, various neural network approaches were used, among which the use of neural networks with adaptive resonance theory; cellular, multilayer perceptron; oscillatory, pulse-coupled, based on probabilistic models; radial basis functions; and the Kohonen self-organizing maps, as described by [46]. Next, some developments and their results are presented, for image recognition in the field of plant species and elements, whose review and analysis allowed the development of the research objective methodological proposal.

Reference [47] developed an artificial vision system for classifying coffee fruits into 11 categories according to their state of maturity, characterized from the shape, size, color, and texture of the fruit. The final set of characteristics was evaluated with three classification techniques, namely: Bayesian, neural networks, and fuzzy clustering. The classification errors were 5.43% for the Bayesian classifier, 7.46% for the neural networks, and 19.46% for the fuzzy clustering. The type of artificial neural network corresponded to a multilayer perceptron with Levenberg-Marquardt learning algorithm.

Visual inspection by humans of wood defects is within a 75–85% accuracy range [48]. developed an artificial neural classifier to recognize and differentiate between seven types of defects in the wood corresponding to the so-called buttons. For this purpose, a set of images with the aforementioned defects, using two-dimensional Gabor filters, reducing the dimensionality of the filter by means of the principal component analysis method. Characteristic extraction was carried out with an artificial neural network, which corresponded to a multilayer perceptron trained with a resilient backpropagation learning algorithm. The performance of the neural network evaluated with the linear correlation coefficient was 83.91%. In the same field as the previous development, [49] identified both the gnarled cylinder and the defect-free zone, in X-ray computed tomography images for pruned radiata pine logs (*Pinus radiata* D. Don), using a supervised classification method based on artificial neural networks. The network used corresponded to a multilayer perceptron, which for thematic maps and for filtered thematic maps had a precision of 92.7% and 96.3% in the identification and separation of the node, evaluated with the linear correlation coefficient.

Researchers concerned about the increase in websites proposing the illicit consumption of cannabis propose an algorithm for the recognition of images with contents of marijuana plants, for which they developed an algorithm based on AdaBoost [50]. For this research, they used 1197 images of marijuana plants and 1821 images from other plants, among which they randomly chose 100 and 300 images, respectively, for training. The images were normalized using bilinear interpolation, whose sizes corresponded to 360 pixels. The results obtained showed a recognition of the cannabis images of 87.5%, higher than other developments carried out, which reached values of 85.3% and 83.5% [50].

Reference [51] report the implementation of a prototype system that performs the classification, by degree of maturity, of Arazá (*Eugenia stipitata* Mc Vaugh) using artificial neural networks. The prototype system consisted of an image acquisition system and a graphical user interface called SISCA, while the artificial neural network corresponded to a multilayer network with backpropagation learning rule that allowed to classify six stages of maturity of the Arazá fruit into RGB color space. The results obtained with the prototype were compared with the results obtained by the classification method of both the laboratory and the collector. The performance of the neural network evaluated with the linear correlation coefficient was 90.32% in the classification of fruit by state of maturity, and the time to perform this classification was reduced by 90%.

Reference [52] propose an interactive system that allows recognizing images of flowers captured from various digital cameras (SONY T9, Canon IS 860, and NIKON S1). For this purpose, they carried out experiments in two databases, one to recognize 24 species of flowers (348 images) and another to recognize 102 species of flowers (612 images), and the proposed system consisted of three main phases, where they make the extraction of the image region of interest, extraction of image features, and image recognition. The results showed recognitions above 90%, superior to other developments found in the literature [52]. In the same field of flower image recognition, [53] developed a flower recognition system doing image processing. For this purpose, the color and border characteristics of the images are used in classifying the flowers. Histograms were used to derive the characteristics of the colors red, green, and blue, in addition to hue and saturation. The characteristics of the edges were made from the application of the Hu's seven-moment algorithm, and the classification of the flowers was carried out using the k-nearest neighbor. The experimentation was carried out on ten species of flowers, with 50 images per species, and 100 images were tested, obtaining a precision of 80%. The researchers note that for two species of flowers, the precision was 100%.

When observing difficulty in a high number of students in the botanical systematics area to classify plant species, [54] developed an intelligent tutor that allows to deliver information, develop practices, and strengthen learning on the topics of the aforementioned subject. The pedagogical tool was called "Mobile Intelligent Tutor for Classification of Plants," and it was developed as an application for cell phones with Android operating system and was based on the use of artificial neural networks to classify the plant.

Reference [55] developed a robust and versatile application for image recognition, in which a neural network was trained to identify tree leaves. In this work, the images were previously processed using OpenCV libraries for Visual Studio software and C++ programming language. To capture the images, he used a cell phone creating an application for Android. The images were transformed from RGB to grayscale by adjusting the weights of red, green, and blue with comparisons of human eye sensitivity visual observations. The image treatment process continued with filtering applying Gaussian blur, binarization, and edge detection using Canny's algorithm [56]. For the extraction of characteristics, he used a backpropagation neural network, which had a performance evaluated with a linear correlation coefficient of 80%.

From the shape function (calculation of the perimeter, length of the major and minor axes, calculation of the area, characteristics of the shape from edge detection techniques), color function (color space such as HSI and HSV for base color classification, calculated from the mean and standard deviation), and the texture function (using the gray-level co-occurrence matrix – GLCM), [57] propose an algorithm for the classification of five species of fruits in which their classification is carried out with an artificial neural network trained to compare 38 parameters of shape, color, and texture. The artificial neural network corresponds to a multilayer perceptron with 2 hidden layers (10 and 5 hidden neurons, respectively, in each hidden layer) and was trained with a set of 100 images. The precision in the estimation was in the range of 92–96% per fruit species.

Reference [58] carried out the recognition of the damages caused by pests in the cultivation of *Begonia semperflorens* (sugar flower) using a multilayer net with supervised training. For this purpose, they made a capture of images using a drone, which received an initial treatment with a morphological dilation filter to increase their pixelation that allowed to identify the perforations in the leaves, and later the images were smoothed using a filter based on Gaussian blur. Finally, it was complemented with two processes, one of them with filtering to obtain images only of the leaves with adjustments in luminosity, saturation, and tones, while the other corresponded to obtaining images in black and white through binarization of the set threshold. This project was developed through the Visual Express platform using Aforge.net free library, and a multilayer neural network was used whose performance showed an error of 9.8% and a linear correlation coefficient of 90.2%.

Reference [59] proposed a backpropagation multilayer artificial neural network with supervised learning trained with four different learning algorithms independently, to commercially classify Royal Gala apples. Its set of images, based on 30 apples with daily monitoring for 35 days, was made up of 4200 photographs in whose processing and resizing to the 256 × 456 size were converted to RGB, HSV, and Lab color models, in order to extract the values of the mean, variance, and standard deviation. The best precision result, based on the linear correlation coefficient, corresponded to the stepped gradient learning algorithm conjugated with a value of 67.7%.

Using a combination of an image processing method (watershed algorithm) and a predictive statistical analysis model, [60] developed a tool to evaluate the number of flower buds in the inflorescence of *Vitis vinifera* L. (the cultivar Cardinal). The research was developed in a viticulture and processed for training 80 images of inflorescence of the vine, and the results in the test with images of 238 flowers were correlated with the results of the manual counting of 301 flowers, observing a strong correlation in the comparison of both methods ($R^2 = 0.98$). Also, for inflorescences in vineyards, [61] develop an investigation in which an image recognition tool performs the counting of inflorescences and non-inflorescences in flowers of grapevines. The classification of the inflorescence or non-inflorescence state was carried out using an artificial neural network, in this case being a fully convolutional network (FCN). The results showed an ability to recount around 80% of the inflorescences in the processed images, with a classification precision of 70%. Likewise, [62] used in a flower image recognition system and different convolutional neural network architectures trained with a set of 9500 images. The best recognition performance reached 97.78%.

Reference [63] implemented a tool called "OrnaNet" to predict the presence of pathogens in the production of ornamental plants using convolutional neural networks, whose registered users have access to the web portal. To operate the tool, the user uploads a photograph of the injured plant to it and the neural network predicts the type of pathogen as well as the damage caused and its location. In a similar field of research and using a set with 8189 images of flowers belonging to 102 species, [64] developed a recognition system for the Oxford 102 flower data set. In this research the processing technique of the images was raised in four stages: (1) image enhancement by cropping and modifying them; (2) image segmentation to separate the object from the background; (3) extraction of characteristics such as color (using HSV descriptor), texture (using gray-level co-occurrence matrix), and shape (using invariant moments); and (4) classification process (using an artificial neural network). In this investigation, the neural network consisted of a backpropagation multilayer perceptron with 20 input parameters, 2 hidden layers (with 100 and 102 neurons, respectively), and a final layer to classify the flowers of the 102 species. The neural network performs the classification with an accuracy of 81.19%.

One of the problems detected in the Peruvian agricultural sector for blueberry crops is the difficulty of detecting diseases and pests despite the use of drones and observations with human vision by farmers. Due to the aforementioned situation, [65] built an expert system to detect diseases and pests in blueberry leaves. The expert system, from the processing and analysis of images in blueberry crops, used machine learning for the use of learning algorithms as classifiers, this being the support vector machine and deep convolutional artificial neural networks. Regarding neural networks, they obtained a performance evaluated with the linear correlation coefficient of 85.6%, which surpassed the other classifiers used (support vector machine, 82.4%, and random forest, 62%).

Reference [66] develop a classification system for medicinal plant leaves and allow two task groups to be carried out. The first task classifies the plant leaf between calendula leaf and tisane leaf, while the second task is aimed at classifying the usefulness of the leaf according to its condition, defining whether it is useful for oil extraction, herbal tea making, or leaf scrap. In the developed system, within the image processing it is developed by means of segmentation algorithms using k-means, and the extraction of the characteristics is carried out by means of trained neural networks, which correspond to multilayer and convolutional perceptron. The system was built on a Raspberry Pi 3 embedded system, which proved to be easy to use and low cost for its test users.

Reference [67] proposed a training set that at the time of the investigation contained 38,409 images of 60 fruits and has subsequently been updated with more new fruit images. They trained a convolutional neural network for the classification of fruits reaching a performance of 96.3%. On the other hand, [68] trained a deep convolutional neural network for the classification of 13 categories of fruits, namely: banana, cherry, strawberry, lemon, mandarin, mango, apple, blackberry, orange, papaya, pear, pineapple, and grape. The capture of 13,516 images was performed using digital devices (13-megapixel camera) and digital downloads with good resolution available on the Internet, which were processed with image management tools (Photoshop) and tagged for identification. In more recent studies, for the identification of images of bananas and apples, [69] propose an image processing technique in which the classification methods were the k-nearest neighbor algorithm (KNN) and support vector machine (SVM), obtaining with the SVM a 100% recognition of the images.

Reference [70] propose the recognition of large-flowered chrysanthemum in cultivars, using deep learning models based on two artificial neural network architectures which are deep convolutional neural networks (DCNN). For the training of neural networks, a set with 14,000 images from 103 cultivars was made using an automatic image acquisition device to photograph from different perspective planes in the said cultivars. The architectures of the neural networks used corresponded to VGG16 (very deep convolutional networks for large-scale image recognition) and ResNet50 (deep residual learning networks for image recognition), reported in 2014 and 2016, respectively. Researchers report a performance greater than 98%. Along the same lines, [71] using deep convolutional neural networks propose a fruit recognition system. The networks were trained with 17,823 images corresponding to 360 fruits grouped in 25 categories and the researchers report a performance in recognition of 99.79%.

Reference [72] describe the techniques and processes used in a system for the identification and classification of diseases in potato crops using image processing that allowed determining and detecting the phases in diseases caused by three types of pests present in potato crops, namely: alternariosis (*Alternaria solani*), late blight (*Phytophthora infestans*), and viruses (PVS). The development was carried out through five phases, which began with the acquisition of images of diseased potato leaves and ended with the classification of the disease caused by the type of pest described previously, using an artificial neural network. The report does not mention the type of neural network used or the results obtained in terms of the performance of the said neural network.

In conclusion, the literature review shows various applications for the recognition of images of fruits, seeds, leaves, and flowers, including inflorescences in order to determine their quality for a specific purpose such as harvest or production, as well as the determination of the state of maturity in fruits. However, applications for inflorescences in marijuana plants have not yet been developed, which leads to a fertile field of technological development. It is noted that the main applications have been developed using neural networks as classifiers, mainly deep learning, being the most used typologies of convolutional networks. This conclusion is consistent with recent reviews [73, 74].

3 Methodological Proposal

The biological description and identification of the plant anatomical elements of the plant species, the theoretical foundation on computer vision, and the review of reports related to developments for similar applications in other plant species lead to proposing a methodology for the identification of maturity states in *Cannabis* sp. as a future contribution in decision-making for the optimum harvest time in Colombian crops. According to the literature reviewed, the methodological proposal establishes procedures for data collection, application design, and training of a convolutional artificial neural network as a tool for the extraction of characteristics in inflorescence photographs.

3.1 Methodology for Capturing Images and Database

In this study, as reported by [75], a photographic record was made in a medicinal cannabis-controlled environment crop with homogeneous sowing of the Charlotte's Angel variety with a predominance of *Cannabis sativa* located in the Miravalle Village (Yumbo Municipality, Valle del Cauca Department, Colombia), at an altitude between 1600 and 1700 masl. As a technological tool, a mobile device was used that has an 8-megapixel resolution digital camera to which a 120× magnification lens was adapted to enlarge the image and register the trichomes.

The photographic record was made at a distance of 3 cm from the surface of the trichomes, and a database was created with 270 photographs of which, according to the indications of the scientific staff of the field, 90 photographs correspond to trichomes in immature state (predominance of translucent color), 90 photographs correspond to a mature state (predominance of white color), and 90 photographs correspond to an overripe state (predominance of amber color).

Finally, a preprocessing of the images was carried out by cutting each image in order to eliminate out-of-focus areas from the peripheries of the photograph and discard elements other than the trichomes. This procedure was carried out using a computational code written in the C++ programming language and OpenCv libraries, compatible for the Microsoft Visual Studio integrated development environment (in its free version: Visual C# Express Edition 2019). Figure 5 shows an aspect of the cannabis cultivation where the sample was made for the conformation of the photographic record and an aspect of the final morphology of the preprocessed image of the trichome.

Fig. 5 Aspect of the controlled environment cultivation of cannabis (left) and final appearance of a trichome image, which was preprocessed (right). (Own Elaboration)

3.2 Methodology for the Development of an Application to Recognize the Optimal Harvest Time of the Inflorescences of **Cannabis** *sp.*

In this study, a waterfall model is proposed for software development in accordance with [76], consisting of the following stages: (1) definition of requirements, (2) software and system design, (3) implementation and unit testing, (4) system integration and testing, and (5) operation and maintenance. In this report, the authors focus on the first two steps.

Definition of Requirements

Based on [6], the functional requirements of the application are defined as (1) obtaining the image, (2) identification of trichomes, (3) trichome count, (4) calculation of the proportion of trichomes, and (5) report generation. As nonfunctional requirements, memory freeing and storage space freeing are defined.

Software and System Design

At the design stage, the user is responsible for registering the photograph and deciding whether to keep the information. The system is in charge of processing the image; obtaining, comparing, and classifying the characteristics; and presenting the results. For the development of the application, the training of a convolutional artificial neural network is proposed and RGB is proposed as a color model. Figure 6 shows the general activity of the proposed application, while Fig. 7 shows the activities to be carried out by the neural network.

Fig. 6 Flow diagram for
the general activity of the
application using RGB as a
color model. (Own
Elaboration)

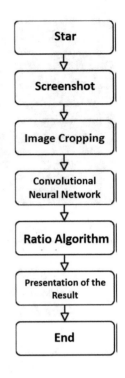

4 Exploration in Image Recognition for Inflorescences of *Cannabis* sp. Using Artificial Neural Networks

In addition to the proposal of a methodology to identify on images the state of maturity in inflorescences of *Cannabis* sp., this study explores the use of neural networks with different typologies to the convolutional neural network. The literature reports for the identification of images the use of neural networks of both multilayer perceptron and radial basis functions. For this reason, this study explores the use of these networks, and the possibility of evaluating the feasibility of implementing the methodological proposal and its reliability in the main objective.

4.1 Construction of the Artificial Neural Network Training Set

For the training of artificial neural networks, a database made up of patterns or information vectors is prepared, which allows the prediction model to recognize the relationships that exist between the input neurons (variables) that lead to the result of the neuron to predict (output variable). For this purpose, based on the photographic record of *Cannabis* sp. inflorescences carried out by [75], a selection of relevant input variables is made in the problem of the maturity state of the aforementioned inflorescences.

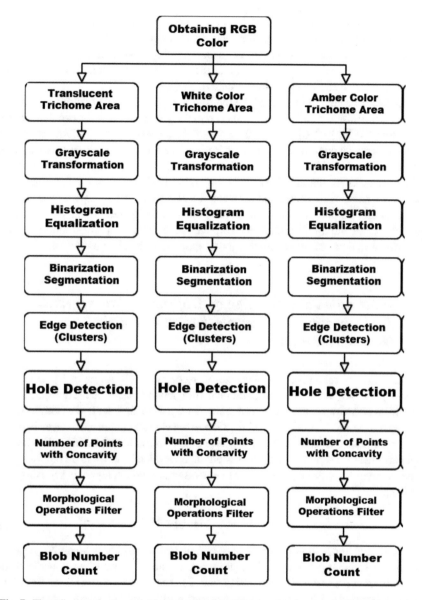

Fig. 7 Flow diagram for the activities to be developed by the neural network. (Own Elaboration)

Based on [77–80], a simulation of an observation process is carried out by a panel of experts in such a way that the variation of the intensity of the color and the variation of the state of maturity respond to random responses. It is considered that an expert could objectively determine the predominant color that the maturity condition infers, while responses on the variation in intensity of the predominant color and the variation in the condition of the maturity state could correspond to subjective assessments which may be represented by random numbers. The photographic record is evaluated, and for each photograph, it is subjected to "objective judgments" about the predominant color in the inflorescence and its state of maturity, and to "subjective judgments" about the intensity of the predominant color in the inflorescence and progress in the state of maturity.

The input variables considered are the following: predominant color in the inflorescence (translucent, white, and amber, represented as a mutually exclusive class variable: [1 0 0], [0 1 0], and [0 0 1]) and the intensity of the predominant color (represented as a continuous variable between 0 and 1). As an output variable, this information for each pattern is associated with the progress of the maturity stage (represented as a continuous variable and rated between 0 and 1 for immature, between 1 and 2 for mature, and between 2 and 3 for overripe). After determining the input and output variables, the development of the general database is carried out with which the training set is determined, composed of 210 information vectors (each information vector was made up of 4 input values and an output value).

4.2 Elaboration and Training of Artificial Neural Networks for Estimating the Maturity State of Cannabis sp. Inflorescences

The training set mentioned in Sect. 4.1 is used, in which the input variables are related to the output variable, an artificial neural network proposal is elaborated, and a multilayer artificial neural architecture fed forward and with methodology of backward training/learning, called feedfoward-backpropagation multilayer perceptron and whose characteristics have been defined in the literature [81]. The multilayer architecture is proposed with an input layer, hidden layers, and an output layer, and the upper limit of the number S of neurons in the hidden layers is set using Eq. (1) [82]:

$$S < \left(N_{\text{DATA}} - 1\right) / \left(n + 3\right) \tag{1}$$

where S is the upper limit of the number of neurons in the hidden layer, N_{DATA} is the number of training patterns (1350), and n is the number of variables considered in the problem to be solved (inputs and output, 5). According to Eq. (1), the limit S is set as 168 neurons in the hidden layer; however, only 50 hidden neurons are required, which are distributed in 3 layers (20-20-10). The connections between neurons in the layers are represented by activation functions. Thus, the connection between neurons from the input layer to the hidden layers and between them is made through the sigmoid function. For its part, the connection between the last hidden layer and the neuron to estimate located in the output layer is made using the linear function, in such a way that the comparison of the estimate and the true result is allowed as described in the supervised training.

The training procedure for the respective set was performed with six conformations of artificial neural network topologies corresponding to six backpropagation training methods. The said conformations were evaluated from the linear correlation coefficient R, a statistic used to evaluate the quality of the performance of the supervised artificial neural networks. For the computational operation of the training procedure, an algorithm was coded using the M programming language, typical of the MATLAB® software tool for the Windows® platform. The code uses the library contained in the neural network toolbox of the same software, which allows implementing the models of the type of network that has been described.

The 210 information vectors of the training set were categorized into three subsets that formed the same number of phases in the neural network training: (1) learning, where the synaptic weights are configured in the connections between neurons; (2) test, with which it was determined you should stop training and optimizing the structure of the neural network and the specifications of the internal model, according to the learning method that uses the backpropagation algorithm; and (3) for validation, where the ability to generalize the model was tested for the range of information that was used for calibration. From the performance indicator, the estimates of the neural networks and the real outputs of the database were evaluated, in the subsets of each training phase (learning, test, validation, and additionally simulation with the total data, in which 60, 20, 20, and 100% of the data, respectively, were considered and grouped by random division of the data).

The results of this process showed that for this case study, the architecture with 3 hidden layers of 50 neurons distributed respectively in 20-20-10 neurons in each hidden layer, and for the 6 learning methods, namely, Levenberg-Marquardt, Quasi-Newton of BFGS Algorithm (Broyden-Fletcher-Goldfarb-Shanno), Resilient Backpropagation, Scaled Conjugate Gradient, One-Step Secant, and Descending Gradient with Variable Learning Rate, obtained adequate and reliable performance ($R \sim 0.95$) in a manner similar between them. As an example, the performance results for the neural network trained with the Levenberg-Marquardt algorithm are shown in Fig. 8. The figure shows that despite the deviation in the estimates, the clustering around the three stages of maturity to be evaluated in the inflorescences is clear, which is what we finally seek to identify.

The grouping of the estimates allows us to infer that the maturity states, observed and evaluated by human vision in this case, are associated with the characteristics of the trichomes of the *Cannabis* sp. inflorescences for these maturity states. This inference on the grouping due to characteristics led in this study to explore another type of neural network, this being a radial basis function network. For this purpose, with the same training set mentioned in Sect. 4.1, a proposal for a radial basis neural network is elaborated, whose characteristics are defined by [83, 84], using as a learning procedure a hybrid algorithm exposed by [85]. The connections between input neurons to the hidden layer are made using the Gaussian function according to [86], and the deviations to the centers are calculated using the radial Euclidean distance defined in [87, 88].

Figure 9 shows the results of the training of the radial basis function neural network, in which it is evident how the estimates are grouped as a cluster around the

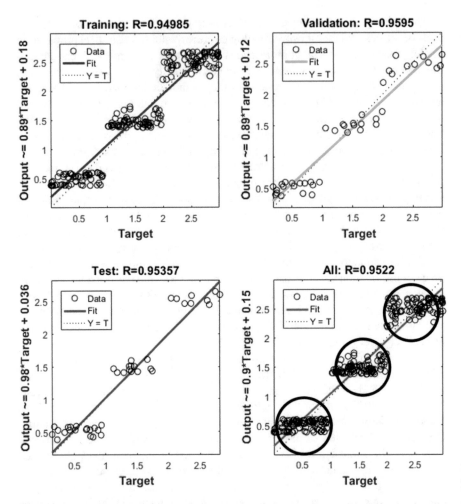

Fig. 8 Performance results of the perceptron multilayer neural network, trained using the Levenberg-Marquardt algorithm, for the estimation of the maturity state in inflorescence trichomes of *Cannabis* sp. (Taken from the result obtained in the software used)

three maturity states considered, which allows inferring that the trichomes of the inflorescences of the crop have similar characteristics in each stage of maturity, and this can be extracted in the case of using specialized identification tools for this purpose, as is the case for convolutional neural networks. In addition, the observer-evaluator of the images has the disadvantages of not always being available and of his subjective judgment, and, the trichomes in different colors (indicator of the state of maturity) are mixed in the inflorescences, providing the process of judgment and observation a greater degree of difficulty and uncertainty.

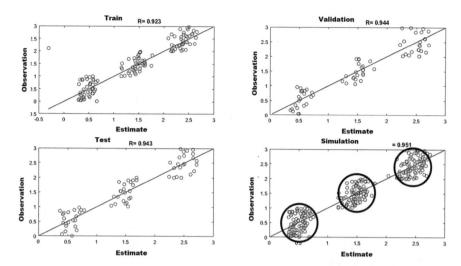

Fig. 9 Performance results of the radial basis function neural network, trained for the estimation of the maturity state in inflorescence trichomes of *Cannabis* sp. (Taken from the result obtained in the software used)

5 Conclusions and Future Agenda

A methodological proposal to develop an application that allows to identify the opportune harvest time in inflorescences of medicinal cannabis is described. The opportune moment is defined as that in which the inflorescences exhibit characteristics corresponding to a predominant presence of white trichomes.

The literature review shows the realization of various developments based on images of plant species to identify pathologies in fruits and leaves, classify fruits, etc., with an adequate degree of certainty greater than the capacity margin of human vision. This led to the methodological proposal of the study described in this chapter being based on the identification of a photographic record of inflorescences of medicinal cannabis and specifically on the extraction of characteristics related to trichomes, among them the predominance of three colors possible (translucent, white, and amber) as well as the proportion of the predominant colors, which corresponds to the main criterion to classify the degree of maturity.

As an advance of the methodological proposal, a photographic record of 210 images was made, whose images were pretreated to limit their content and focus on them the unique identification of trichomes. The content of the images made it possible to form a database that shows the degrees of maturity of the inflorescences, in such a way that an adequate set is provided to apply characteristic extraction tools.

The literature review also showed that in computational vision used in the identification of images and for problems similar to the present study, the feature extraction tools correspond to artificial neural networks, the most common of which are perceptron multilayer networks, networks of radial basis functions, and convolutional networks. In the methodological proposal of this study, the use of a convolutional neural network is proposed as a tool for extracting the characteristics associated with a degree of maturity.

To explore the viability and reliability of the proposed methodology, a process of identification of the maturity states is simulated based on the objectivity and subjectivity of the identification of the said maturity states, in such a way that they could be generalized by artificial neural networks. For this purpose, initially a multilayer perceptron artificial neural network with training based on the backpropagation algorithm was used. The network for the plotted purpose groups the responses in three maturity groups to predict (identify for the image), which allowed us to infer that a network that predicts around the grouped responses was used to confirm this behavior.

For this purpose, a network of radial basis functions was trained, in which the estimates show a cluster behavior for each maturity stage, which allowed to confirm that a neural network specialized in extracting characteristics and grouping them into representative clusters is more appropriate, as in the case of convolutional neural networks. This makes it possible to state that the proposal to use the convolutional neural network as a feature extraction tool is the correct one.

From the results obtained in the exploration with common networks, it could be inferred that the methodological proposal with convolutional networks is viable and reliable for the development of an application that evaluates image records with the trichomes of the inflorescences and allows making decisions about the appropriate moment of the harvest. The culmination of the application development then becomes a future agenda which may include exploration with the inclusion of other feature extraction tools such as other types of specialized networks.

The reasons stated in the previous paragraphs within this section allow us to conclude that artificial vision will contribute to the development of methodologies that allow determining the opportune moment to harvest the inflorescences of medicinal cannabis in crops located in Colombia.

References

1. INFAIMON Homepage, Historia de la visión artificial: así ha evolucionado esta tecnología. [R]evolución artificial (enero 29, 2020), https://blog.infaimon.com/historia-vision-artificial/. Last accessed 22 Oct 2020
2. G. Lio, J. Yang, Z. Li, Content-based image retrieval using computational visual attention model. Pattern Recogn. 8(48), 2554–2566 (2015)
3. I. Berzal, *Desarrollo de algoritmos de procesamiento de imágenes con VTK* (Universidad Politécnica de Madrid, Madrid, 2004)
4. A. Broggi, S. Cattani, P. Medici, P. Zani, Applications of computer vision to vehicles: An extreme test, in *Machine Learning for Computer Vision*, vol. 411, (Springer, Berlin, Heidelberg, 2013), pp. 215–250
5. K. Hatou, A. Pamungkas, T. Morimoto, Image processing by artificial neural networks for stress diagnosis of tomato. IFAC Proc. 44(1), 1768–1772 (2011)

6. E. Sánchez, *Aplicación móvil para el conteo automático e identificador preliminar de colonias de bacterias mediante reconocimiento de patrones* (Instituto Politécnico Nacional, Zacatecas, 2018)
7. E. Rosenthal, D. Downs, *Marijuana Harvest: How to Maximize Quality and Yield in Your Cannabis Garden* (ZLibrary, 2017)
8. H. Kholer, *Kholer's Medizinal-Pflazen*. 3rd vol (Franz Eugen Kholer Editorial, Berlin, 1897)
9. E. Small, *Cannabis. A Complete Guide*, 1st edn. (CRC Press, Ottawa, 2017)
10. D. Jin, S. Jin, J. Chen, Cannabis indoor growing conditions, management practices, and post-harvest treatment: A review. Am. J. Plant Sci. **10**, 925–946 (2019)
11. S. Rimon, S. Duchin, N. Bernstein, R. Kamenestky, Architecture and florogenesis in female Cannabis sativa plants. Front. Plant Sci. **10**, 1–10 (2019)
12. E. Kim, P. Mahlberg, Secretory vesicle formation in the secretory cavity of glandular trichomes of Cannabis sativa L. (Cannabaceae). Mol Cells **15**(3), 387–395 (2003)
13. C. Hammond, P. Mahlberg, Morphogenesis of capitate glandular hairs of Cannabis sativa L (Cannabaceae). Am. J. Bot. **64**, 1023–1031 (1977)
14. E. Small, S. Naraine, Size matters: Evolution of large drug-secretion resin glands in elite pharmaceutical strains of Cannabis sativa (marijuana). Genet. Resour. Crop. Evol. **63**, 349–359 (2016)
15. V. Raman, H. Lata, S. Chandra, I. Khan, M. ElSohly, Morpho-anatomy of marijuana (Cannabis sativa L.), in *Cannabis sativa L. Botany and Biotechnology*, ed. by S. Chandra, H. Lata, M. ElSohly, (Springer, Heidelberg, 2017), pp. 123–136
16. R. Wills, B. McGlasson, D. Graham, D. Joyce, *Postharvest: An Introduction to the Physiology and Handing of Fruit, Vegetables and Ornamentals* (CABI Publishing, Wallingford, 1998)
17. G. Seymour, J. Taylor, G. Tucker, *Biochemistry of Fruit Ripening* (Chapman and Hall, London, 1993)
18. A. López, *Manual para la preparación y venta de frutas y hortalizas. Del campo al mercado* (FAO, Roma, 2003)
19. J. González, J. Moral, *Recolección, almacenamiento, y transporte de flores y hortalizas* (IC Editorial, Málaga, 2018)
20. R. Dos Santos, L. Arge, S. Costa, N. Machado, P. De Mello-Farias, C. Rombaldi, A. De Oliveira, Genetic regulation and the impact of omics in fruit ripening. Plant Omics J. **8**(2), 78–88 (2015)
21. M. Martínez, R. Morales, I. Tejacal, M. Cortés, Y. Palomino, G. López, Poscosecha de frutos: maduración y cambios bioquímicos. Rev. Mex. Cienc. Agríc. **19**(12), 4075–4087 (2017)
22. J. Hernández-Orallo, F. Martínez-Plumed, U. Schmid, M. Siebers, D. Dowe, Computer models intelligence test problems: Progress and implications. Artif. Intell. **230**, 74–107 (2016)
23. C. Berrocal, I. Lofgren, K. Lundgren, N. Gorander, C. Halldén, Characterisation of bending cracks in R/FRC using image analysis. Cem. Concr. Res. **90**, 104–116 (2016)
24. I. Michalska-Pozoga, R. Tomkowski, T. Rydzkowski, V. Kumar, Towards the usage of image analysis technique to measure particles size and composition in wood-polymer composites. Ind. Crop. Prod. **92**, 149–156 (2016)
25. A. Bouchet, P. Alonso, I. Pastore, S. Montes, I. Díaz, Fuzzy mathematical morphology for color images defined by fuzzy preference colors. Pattern Recogn. **60**, 720–733 (2016)
26. S. Ngan, A unified representation of intuitionistic fuzzy sets, hesitant fuzzy sets and generalized hesitant fuzzy sets based on their u-maps. Expert Syst. Appl. **69**, 257–276 (2017)
27. G. Miranda, J. Felipe, Computer-aided diagnosis system based on fuzzy logic for breast cancer categorization. Comput. Biol. Med. **64**, 334–346 (2015)
28. V. De Albuquerque, A. De Alexandria, P. Cortez, J. Tavares, Evaluation of multilayer perceptron and self-organizing map neural network topologies applied on microstructure segmentation from metallographic images. NDT E Int. **42**(7), 644–651 (2009)
29. A. Mashaly, A. Alazba, MLP and MLR models for instantaneous thermal efficiency prediction of solar still under hyper-arid environment. Comput. Electron. Agric. **122**, 146–155 (2016)

30. Z. Arjmandzadeh, M. Safi, A. Nazemi, A new neural network model for solving random interval linear programming problems. Neural Netw. **89**, 11–18 (2017)
31. C. Huang, H. Li, W. Li, Q. Wu, L. Xu, Store classification using Text-Exemplar-Similarity and Hypotheses-Weighted-CNN. J. Vis. Commun. Image Represent. **44**, 21–28 (2017)
32. K. Apostolidis, L. Hadjileontiadis, Swarm decomposition: A novel signal analysis using swarm intelligence. Signal Process. **132**, 40–50 (2017)
33. M. Mavrovouniotis, C. Li, S. Yang, A survey of swarm intelligence for dynamic optimization: Algorithms and applications. Swarm Evol. Comput. **33**, 1–17 (2017)
34. D. Kumar, K. Mishra, Portfolio optimization using novel co-variance guided Artificial Bee Colony algorithm. Swarm Evol. Comput. **33**, 119–130 (2017)
35. B. Schmidt, A. Al-Fuqaha, A. Gupta, D. Kountanis, Optimizing an artificial immune system algorithm in support of flow-based internet traffic classification. Appl. Soft Comput. **54**, 1–22 (2017)
36. A. Hatata, E. Abd-Raboh, B. Sedhom, Proposed Sandia frequency shift for anti-islanding detection method based on artificial immune system. Alex. Eng. J. **57**(1), 235–245 (2018)
37. R. Kuo, Y. Tseng, Z. Chen, Integration of fuzzy neural network and artificial immune system-based back-propagation neural network for sales forecasting using qualitative and quantitative data. J. Intell. Manuf. **27**, 1191–1207 (2016)
38. U. Greeshma, S. Annalakshmi, Artificial neural network (research paper on basics of ANN). Int. J. Sci. Eng. Res. **6**(4), 110–115 (2015)
39. A. Markopoulos, S. Gergiopoulos, D. Manolakos, On the use of back propagation and radial basis function neural networks in surface roughness prediction. J. Ind. Eng. Int. **12**, 389–400 (2016)
40. J.-L. Ramírez-Arias, A. Rubiano-Fonseca, R. Jiménez-Moreno, Object recognition through artificial intelligence techniques. Rev. Fac. Ing. **29**(54), e10734, 1–18 (2020)
41. I. Goodfellow, Y. Bengio, A. Courville, *Deep Learning* (MIT Press, Cambridge, 2016)
42. A. Krizhevsky, I. Sutskever, G. Hinton, ImageNet classification with deep convolutional neural networks. Adv. Neural Inf. Proces. Syst. **25**(2), 1097–1105 (2012)
43. R. Yamashita, M. Nishio, R. Do, K. Togashi, Convolutional neural networks: An overview and application in radiology. Insights Imaging **9**(4), 1–19 (2018)
44. M. Alom, T. Taha, C. Yakopcic, S. Westberg, P. Sidike, M. Nasrin, M. Hasan, B. Essen, A. Awwal, V. Asari, A state-of-the-art survey on deep learning theory and architectures. Electronics **8**(292), 2–67 (2019)
45. W. Rawat, Z. Wang, Deep convolutional neural networks for image classification: A comprehensive review. Neural Comput. **29**, 2352–2449 (2017)
46. O. Abiodun, A. Jantan, E. Omolara, K. Dada, N. Mohamed, State-of-the-art in artificial neural network applications: A survey. Heliyon **4**(11), e00938 (2018)
47. Z. Sandoval, *Caracterización y clasificación de café cereza usando visión artificial* (Universidad Nacional de Colombia Sede Manizales, Manizales, 2005)
48. G. Ramírez, M. Chacón, Clasificación de defectos en madera utilizando redes neuronales artificiales. Comput. Sist. **9**(1), 17–27 (2005)
49. G. Rojas-Espinoza, O. Ortiz-Iribarren, Identificación del cilindro nudoso en imágenes TC de trozas podadas de pinus radiata utilizando redes neuronales artificiales. Maderas Cienc. Tecnol. **12**(3), 229–239 (2010)
50. L. Xie, X. Li, X. Zhang, W. Hu, J.Z. Wang, Boosted cannabis image recognition, in 2008 19th International Conference on Pattern Recognition, Tampa, FL (2008), pp. 1–4, https://doi.org/10.1109/ICPR.2008.4761592
51. L. España, C. Camacho, L. Marín, Sistema prototipo para clasificación de Eugenia stipitata por grado de madurez mediante redes neuronales artificiales. Ing. Amazon. **3**(2), 119–127 (2010)
52. T.H. Hsu, C.S. Lee, L.H. Chen, An interactive flower image recognition system. Multimed. Tools Appl. **53**, 53–73 (2011). https://doi.org/10.1007/s11042-010-0490-6
53. T. Tiay, P. Benyaphaichit, P. Riyamongkol, Flower recognition system based on image processing, in 2014 Third ICT International Student Project Conference (ICT-ISPC2014) (2014)

54. T. Roca, *Tutor inteligente móvil para la clasificación de plantas basado en redes neuronales* (Universidad Mayor de San Andrés, La Paz, 2013)
55. P. García, *Reconocimiento de imágenes utilizando redes neuronales artificiales* (Universidad Complutense de Madrid, Madrid, 2013)
56. J. Canny, A computational approach to edge detection. IEEE Trans. Pattern Anal. Mach. Intell. **8**(6), 679–698 (1986)
57. B. Pratap, N. Agarwal, S. Joshi, S. Gupta, Development of ANN based efficient fruit recognition technique. Global J. Comp. Sci. Technol. **14**(5), 1–6 (2014)
58. C. Cáceres, O. Ramos, D. Amaya, Procesamiento de imágenes para reconocimiento de daños causados por plagas en el cultivo de Begonia semperflorens (flor de azúcar). Acta Agron. **64**(3), 273–279 (2015)
59. G. Figueredo, Clasificación de la manzana roya gala usando visión artificial y redes neuronales artificiales. Res. Comput. Sci. **114**, 23–32 (2016)
60. R. Benhehaia, D. Khedidja, M.E.M. Bentchikou, Estimation of the flower buttons per inflorescences of grapevine (Vitis vinifera L.) by image auto-assessment processing. Afr. J. Agric. Res. **11**(34), 3203–3209 (2016). https://doi.org/10.5897/AJAR2016.11331
61. R. Rudolph, K. Herzog, R. Töpfer, V. Steinhage, Efficient identification, localization and quantification of grapevine inflorescences in unprepared field images using Fully Convolutional Networks. J. Grapevine Res. Vitis **58**(3), 95–104 (2019)
62. M.V.D. Prasad, B.J. Lakshmamma, A.H. Chandana, K. Komali, M.V.N. Manoja, P.R. Khumar, C.R. Prasad, S. Inthiyaz, P.S. Kiran, An efficient classification of flower images with convolutional neural networks. Int. J. Eng. Technol. **7**(1.1), 384–391 (2018)
63. E. Escobar, *Predicción de agentes patógenos en plantas ornamentales utilizando redes neuronales* (Instituto Tecnológico de Colima, Villa de Álvarez, 2018)
64. H. Almogdady, S. Manaseer, H. Hiary, A flower recognition system based on image processing and neural networks. Int. J. Sci. Technol. Res. **7**(11), 166–173 (2018)
65. C. Sullca, C. Molina, C. Rodríguez, T. Fernández, Detección de enfermedades y plagas en las hojas de arándanos utilizando técnicas de visión artificial. Perspectiv@s Rev. Technol. Inform. **15**(15), 32–39 (2018)
66. R. Gaviria, C. Marín, *Sistema de inspección y clasificación de hojas de plantas medicinales por medio de visión artificial* (Universidad Autónoma de Occidente, Santiago de Cali, 2018)
67. H. Muresan, M. Oltean, Fruit recognition from images using deep learning. Acta Univ. Sapientiae Inform. **10**(1), 26–42 (2018)
68. J. Aguilar-Alvarado, M. Campoverde-Molina, Clasificación de frutas basadas en redes neuronales convolucionales. Polo del Conocimiento **5**(1), 3–22 (2019)
69. P.L. Chithra, M. Henila, Fruits classification using image processing techniques. Int. J. Comput. Sci. Eng. **7**(5), 131–135 (2019)
70. Z. Liu, J. Wang, Y. Tian, S. Dai, Deep learning for image-based large-flowered chrysanthemum cultivar recognition. Plant Methods **15**(146), 1–11 (2019)
71. S. Sakib, Z. Ashrafi, M.A.B. Sidique, Implementation of fruits recognition classifier using convolutional neural network algorithm for observation of accuracies for various hidden layers. arXiv, 1–14 (2019)
72. N. Rosero, J. Cabrera, O. Anrango, M. Yandún, S. Lascano, Detección de enfermedades en cultivos de papa usando procesamiento de imágenes. Rev. Cumbres **6**(1), 43–52 (2020)
73. F. Liu, L. Snetkov, D. Lima, Summary on fruit identification methods: A literature review. Adv. Soc. Sci. Educ. Humanit. Res. **119**, 1629–1633 (2017)
74. J. Naranjo-Torres, M. Mora, R. Hernández-García, R.J. Barrientos, C. Fredes, A. Valenzuela, A review of convolutional neural network applied to fruit image process. Appl. Sci. **10**(3443), 1–31 (2020)
75. A. Palomino, *Exploración para el desarrollo de un aplicativo que permita identificar el momento óptimo de cosecha en inflorescencias de Cannabis Sp para fines medicinales y científicos* (Universidad Nacional de Colombia Sede Palmira, Palmira, 2020)

76. W. Humphrey, *A Discipline for Software Engineering*, 19th edn. (Addison-Wesley, Reading, 1995)
77. J. Parra, Simulación. Rev. Colomb. Estadíst. **3**, 21–50 (1981)
78. P. L'Ecuyer, Random numbers for simulation. Commun. ACM **33**(10), 85–97 (1990)
79. D. DiCarlo, *Random Number Generation: Types and Techniques* (Liberty University, Lynchburg, 2012)
80. G. Izarikova, Process simulation and methods of generating random numbers. Acta Simul. **1**(2), 1–4 (2015)
81. G. Hinton, Connectionist learning procedures. Artif. Intell. **40**(1–3), 185–234 (1988)
82. L. González-Salcedo, J. Gotay-Sardinas, M. Roodschild, A. Will, S. Rodríguez, Optimización en la elaboración de redes neuronales artificiales adaptativas usando una metodología de algoritmo de poda. Ingenio Magno **8**(1), 44–56 (2017)
83. H. Mhaskar, Neural networks for optimal approximation of smooth and analytic functions. Neural Comput. **8**(1), 164–177 (1996)
84. M. Buhmann, *Radial Basis Functions: Theory and Implementations* (Cambridge University Press, Cambridge, 2003)
85. Y. Liao, S. Fang, H. Nuttle, Relaxed conditions for radial-basis function networks to be universal approximators. Neural Netw. **16**, 1019–1028 (2003)
86. C. García, *Redes neuronales de funciones de base radial* (Universidad de La Laguna, La Laguna, 2017)
87. C. Cuadras, Distancia Estadísticas. Estadíst. Española **30**(119), 295–378 (1989)
88. R. Prieto, *Técnicas estadísticas de clasificación, un ejemplo de análisis de clúster* (Universidad Autónoma del Estado de Hidalgo, Pachuca, 2006)

Detection of Brain Tumor Region in MRI Image Through K-Means Clustering Algorithms

Sanjay Kumar ⓘ, Naresh Kumar ⓘ, J. N. Singh, Prashant Johri, and Sanjeev Kumar Singh ⓘ

1 Introduction

The abnormal and uncontrolled growth of brain cells may be described as a tumor. Tumors in any area of the brain may be identified as an irregular buildup of abnormal tissues. Benign tumors are known as non-cancerous and cannot impact the stable ecosystem around them. The gradual process of development of the brain tumor occurs. The malignant tumor is thought to be cancerous and has accelerated growth. With time passing, it leads to a person's death [1]. Many methods are used, but the most widely used method is MRI, for the detection of any anomalies in the body. Information from high-quality pictures is used to study the anatomy of various parts of the body. Two important steps for medical image processing are taken: one is pre-processing and the other is post-processing. In medical imaging techniques, pre-processing of MRI images is an essential process. Pre-processing involves amplification of pictures and reduction of noise. Noise-free and improved MRI image is essential for the execution of any higher-order process. As previously mentioned, many medical image segmentation approaches exist, but K-means clustering algorithms are most commonly used to segment brain tumors. However, several big disadvantages of the K-means clustering algorithms have been reported in a recent paper, for instance the random start of cluster cancroids, noise resistance, increased computational sophistication, and lack of efficacy during complex segmentation of tumors [2]. The primary objective of this article is therefore to develop an enhanced

S. Kumar (✉) · S. K. Singh
Information Technology, Galgotias College of Engineering &Technology, Greater Noida, Uttar Pradesh, India

N. Kumar · J. N. Singh · P. Johri
School of Computing Science & Engineering, Galgotias University, Greater Noida, Uttar Pradesh, India
e-mail: kumar.naresh@galgotiasuniversity.edu.in

© Springer Nature Switzerland AG 2022
P. Johri et al. (eds.), *Trends and Advancements of Image Processing and its Applications*, EAI/Springer Innovations in Communication and Computing, https://doi.org/10.1007/978-3-030-75945-2_11

clustering approach based on the K-means algorithm of the brain tumor section in MRI images. The main concept is to merge K-means clustering to maximize the collection of initial centers for K-medium clusters. Pre-processing is important for the retrieval of simple, precise, and reliable data to better examine anomalies such as the brain tumor in medical imaging. These methods are typically applied to enhance the image consistency and followed by edge detection processes, thresholds, segmentations, and clusters that are part of post-processing. This theoretical approach concentrates mainly on geographic segmentation. Segmentation is not much simpler than supplementary steps throughout image processing with MRI. Segmentation can be used as the initial pace to better interpret medical pictures and to find the variety of elements in the image [3]. For a successful medical examination, the standard MRI images are not ideal, which is why segmentation technique is the most important method used to identify the brain tumor accurately in MRI images. Segmentation strategies are used in every area of the human brain to distinguish between normal and dysfunctional cells [4]. Therefore, for this treatment, tumor identification is critical. In the current stage when the tumor is diagnosed, the lifespan of the individual person affected by the brain tumor can improve. This would make life more stable by 1 or 2 years roughly. Typically tumor cells are categorized as benign and malignant. It's somewhat difficult to detect the malignant tumor massively. For the exact detection of the malignant tumor, a 3D brain and 3D analysis tool are needed. Here, we concentrated in this paper on mass detection of tumors. The measurement tool is Matt laboratory. Since designing and running is fast. Finally, we supply the tumor and its structure with structures [3].

During these days, a vast range of strategies, including cluster, neural networks, flooded logics contouring, regionally dependent segmentation, etc., are being applied (Fig. 1).

Fig. 1 Brain tumor [3]

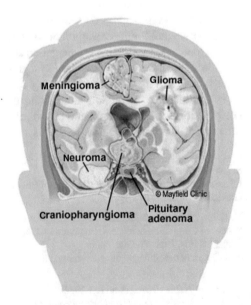

For classification and segmentation of MRIs, it requires many simple and essential procedures such as filtration and border detection to create a better image of MRI. A noiseless MRI image with correct edge detection is critical to the accurate and effective analysis before proceeding to segmentation [5]. This technique has demonstrated that MRI images appear to be distorted by elevated levels of "salt-and-pepper" noise. To produce error-free MRI images, the median filters are used primarily because of the quality of the error and to prevent blurring results. In order to improve the tumor region's boundaries and ensure smooth transitions in the MRI scan, the filtered image is exposed to the hybrid edge preservative. After these foundational steps have been applied, the tumor region's high-intensity pixels are clustered by the use of the K-means clustering and further reinforced by the stream segmentation in order to separate tumor areas and brain areas untouched [5].

2 Literature Review

The findings presented in this method revealed that the watershed segmentation with a contrast technique has also contrasted two morphological wearing away and dilation operators. By implementing watershed segmentation, the areas are separated [1].

After extracting the noise using a median filter, clusters of pixels are formed based on random censor. This methodological approach was based on the clustering technique, primarily on the thresholding by the surface extraction in the MR image. The distance between the clusters and determined with the Euclidean remote procedure. This approach has shown that the instance use of the C-means system is senior and more multifaceted between the K-means and C-means cluster technique [3].

This research paper has been providing a broad distinction between geographic segmentation cluster techniques, plus thresholding bottom approaches. It found the threshold strategy ideal for high-contrast MR images. It also noted that regionally dependent segmentation is suitable for the extraction of objects; however, it can involve kernel collection errors and clustering strategies suited for larger images plus lower degrees of difference [4].

The research before us focused primarily on applying K-means clustering to classify the brain tumor. It listed linear and nonlinear noise reduction filters, as healthy and several edge detection masks. Finally, the MRI image is segmented according to multiple sets of tumor detection clusters [3].

This research article listed the implementation of the segmentation of the watershed, the threshold, and morphological operators for the identification of brain tumors. At the same point, watershed and threshold segmentation have been used to fill trout such as dilatation, erosion, etc. [5].

Segmentation and some morphological techniques to generate MRI images are utilized. The first step was to improve the consistency and then to incorporate morphological methods to diagnose the tumor. In the pre-processing period, a criterion to reach higher pixel values through the sharpening techniques than during the post-processing step. The segmentation of the stream and morphology operators was used after the point. Each technique's findings have been seen separately [6].

The paper describes a tool for brain and tumor segmentation. Three-dimensional data sets have been derived and edema dependent on various characteristics. This new algorithm provided an extension of the previously submitted segmentation (EM) approach used to change the probabilistic brain atlas. This job has been achieved by applying pre- and post-processing of MRI image data sets. The findings are determined by the oncology department based on five separate cases [7].

For the threshold parameters of the MRI images, we used the K-means clustering method in the article. The K-means algorithm starts with the initialization, which is defined for each cluster of average vector iteration. In the second step is specified the class whose pixel vector is nearest to the average vector or each pixel was assigned to that category. The first set of judgment constraints is this pixel range [8].

The study of the cluster or the cluster is a method by which objects are clustered into clusters of related objects. The partitioning of the algorithms has several approaches. The K-means clustering algorithm is one of the most common techniques. We will research about the K-means clustering system and some valuable mathematical instruments that are necessary to develop the proposed method [9].

Various forms of tumors are present. It may be as a mass in the brain or as malignant in the liver. Assume it is a number, so K implies that it is necessary to remove the algorithm from the brain cells. When the MRI image contains noise, it was removed before the K-means method. As input to the K-means and extracting the tumor from the MRI image, the noise-free image is given. And the fuzzy C segmentation implies specific tumor type isolation and threshold of the production of malignant tumors during the removal of the element [10].

3   Proposed Method

The proposed method for segmentation of the brain tumor on MRI images is described in this section. This method is particularly useful for removing some significant K-means classification algorithm shortcomings, such as random cancer cluster initialization and incorrect tumor type estimation. Three types of algorithms are introduced as a post-processing approach for achieving efficient diagnosis of a tumor, such as the DPSO algorithm, the K-means algorithm, and morphologic reconstruction. It allows us to reserve contours of artifacts and reduces noise without knowing the noise form beforehand [9]. For reconstruction, we selected the MRI operator and a morphological closure.

Two K-means and MRI methods allow hybridization with a high degree of precision and less acoustic tumor isolation. First cluster centers, the DPSO algorithm is used and the main steps are described in the proposed procedure. The optimized values for the initial cluster centers can be obtained in this step [10]. The images are separated by a normal algorithm of K-means. The MRI is then added to the image

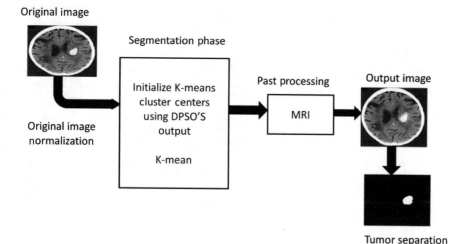

Fig. 2 Different phases of the proposed method

to better isolate the tumor structure. Figure 2 shows a schema that explains the different phases of the process.

For the intended technological algorithm, the simulated code is defined as:

- The only picture has to be normalized in range [0 1].
- Alter the parameter of the DPSO (Table 1).
- Implementation of the DPSO algorithm.
- As a preliminary algorithm, the DPSO implications are old.
- K-means center combined algorithm.
- Run the algorithm of K-means.
- Insert the K-means segmented picture with MRI.

The source image should be noted as having a pretense color to differentiate the region more efficiently. In addition, the MRI's knowledge is focused on the subject of the SEA alternative, imagery, and façade films.

4 K-Means Clustering Algorithm

K-means is one of the unregulated classroom knowledge algorithms. The picture is clustered according to certain attributes by grouping the pixels. We must first specify the number of clusters "k" in the K-means algorithm. Then the middle of the K-cluster is randomly selected [11]. The distance to each cluster center for each pixel is determined. The gap can be simple Euclidean. A single pixel with a distance formula is compared to all cluster centers. The pixel is transferred to a specific cluster with the shortest distance between all clusters. The central one is then

Table 1 Initial parameters of the DPSO

Parameter	Value
V1	0:9
V2	0:9
D1&D2	Accidental
X	2:1
Umin	1:8
Umax	2:8
Existing inhabitants of the group	35
Minimum figure of swarm	5
Maximum numeral of swarm	8
Minimum populace	9
Maximum residents	40
Stagnancy	9
Number of iteration	97

reassessed. Again, all cancroids are likened to any pixel. The loop goes on until the center comes together.

4.1 Mathematical Representation

Calculate the cluster *m* for a particular picture:

$$M = \frac{\sum_{i:c(i)=k} x_i}{N_k}, \quad k = 1,\ldots\ldots,K \tag{1}$$

Compute the aloofness flanked by the bunch centers to every pixel:

$$E(i) = \arg\min x_i - \left| M_k \right|^2, \quad i = 1\ldots\ldots\ldots R \tag{2}$$

Repeat step 2 before convergence of the mean value.

4.2 Algorithm

1. Offer a cluster value not like *k*.
2. Select *k* cluster center arbitrarily.
3. The mean or core of the cluster is determined.
4. Calculate each pixel distance to each center of the cluster.
5. When the detachment is close to the middle, go to the cluster.
6. Go to the next and come together otherwise.

7. Return to the center to estimate.
8. Repeat until the core is not moving.

4.3 Pre-processing

Noise and filters are used in the pre-processing. Any external disruption, including patient movement, improper examination of the region, and machine vibration may result in image deterioration during an acquisition of an MRI image. The composition of a picture is influenced by several different forms of noises. Salty and intersperse noise is one of the major common causes of disruption in the last stage of production. As sound reduces the accuracy of an MRI image, the use of appropriate filters is important to eliminate it. The noise of impulse is a black and pallid pixel or a pixel in the image. A morphological filter or a median filter [11] will suppress the salt-and-pepper noise. Most of the medical images are subject to varying noise levels like deviating from the original data, which is more surprising in image concentration. Without noise, image analysis can be more workable. Linear and nonlinear filters may also be classified. Averaging mask is used in the linear filter, which replaces each pixel by the mean value in its surroundings. In this proposed analysis, limits occupy an important position in addition to treatment except these averaging filters blur an MRI image as the result of sharp transitions on the edges [12]. In order to provide a blur effect to the image, linear filters are used using the correlation and convolution process. Nonlinear spatial filters are based on the pixel ranking in an image region. The classification method is used to organize pixels in down- or upward arrangement. It increases the acuity of an image, decreases the sound content, and retains the limits of the rudiments in a picture.

4.4 Post-processing Comprises

The finishing step involves thresholding, border detection, and segmentation algorithms in the proposed methodology. The threshold is one of the most important segmentation parts of the picture. The threshold is another way to transform a gray picture keen on a dual picture by calculating an effective threshold value. Thresholding can be used to extract useful MRI image information. The best way to divide an image [13] is through this method. Low down doorsill pixels are represented as black and high threshold values as white, the main idea behind threshold segmentation. Limits can be identified as areas in MRI imagery in which the image elements often vary in intensity. Continuity of frontiers is characterized by edges. Border detection is a mathematical method used to enhance image excellence and protects poor object outlines [14]. Clever and Prewitt and Robert border detectors are the most widely used edge operators. Edge operators are not pixel values dependent on their coefficient. The coefficients are calculated from an estimate of small differences. For both horizontal and vertical image directions, edge detectors are usually used. The corium border detector is one of the major detectors that simultaneously notice border and edge direction. It consists primarily of two vertical and horizontal masks. The edge detector coefficients can be determined by a finite formula of the approximation of the difference and modified to our requirement [15].

Nx			Ny		
−1	1	0	1	2	1
−2	2	0	0	0	0
1	1	0	−1	−2	−1

For the measurement in x and y direction of the pixel, the gray edge detector is used. It takes a derivation for horizontal direction between the first and third rows and takes a derivative for vertical direction between the third and first column.

The horizontal axis derivative is shown in Eq. (3).

The vertical axis derivatives are shown in Eq. (4).

$$Hx = (X7 + 2X8 + X9) - (X1 + 2X2 + X3) \tag{3}$$

$$Hy = (X3 + 2X6 + X9) - (X1 + 2X4 + X7) \tag{4}$$

Further, by using this formula, you calculate the image gradient:

$$H = \sqrt{Hx + Hy} \tag{5}$$

Non-maximum deletion is used to remove unacceptable pixel information that cannot be considered part of the edge. The goal of non-maximum control is to dilute boundaries. Complete four suitable directions are used for detecting the edge pixels. Hysteresis thresholds are the process in which two thresholds (top and bottom) are identified. Pixels over the higher doorsill are known as the margin, while the lower threshold value pixels are not identifiable as the margin [15]. Pixels between two threshold values are only considered valid since they are related to the upper threshold value.

At this tip (5 × 5) Gaussian sift with $\sigma = 1.4$

$$B = \underline{1} \begin{bmatrix} 3 & 5 & 4 \\ 4 & 6 & 6 \\ 5 & 5 & 7 \end{bmatrix} \times A$$

A 3*35 Gaussian filter window displays a slightly better performance compared to a different size window.

5   Simulation and Results

This section examines the efficiency of the proposed approach to brain tumor segmentation. Some materials and mathematical methods are shown in the table before such experiments are carried out.

Figure 4 is the MRI image given for the pre-processed and algorithms for K-means. 0.05% of the salt-and-pepper sound was applied and the median filter removed. The K-means algorithm bundles the picture according to those features [16]. Figure 3 is the performance of the five-cluster K-means algorithm. The tumor is removed in the fifth line.

This is why we translate the RGB image into a gray image for a DICOM picture in RGB. We use a kernel of 5 × 5 and use zeroes to reduce the noise in order to add the medium filter. In Fig. 4 are shown padded images (Fig. 5).

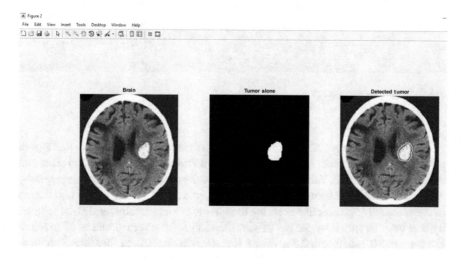

Fig. 3   Output image for pre-processing and K-means for $k = 3$

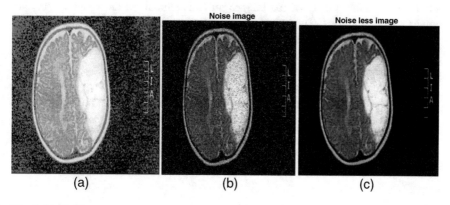

Fig. 4   Padded images

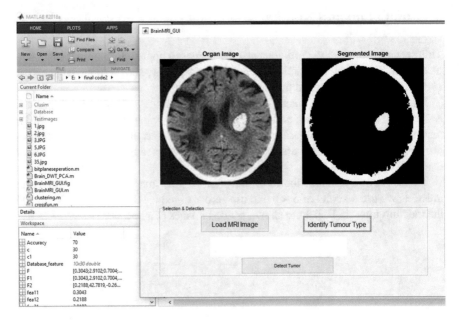

Fig. 5 Cluster K-means result

Various MRI images of the brain are used to assess the validity of the findings in this structured approach. All three pictures contain various structures and sizes and the location of tumors. Various criteria are typically used to determine image quality after algorithms have been used. PSNR, MSE, and SSIM are the image parameters in our research approach that describe the quality of our findings, and also the quality can be interpreted by means of visualization for the measurement of medium filter; we use 0.3 in image, 0.5 in image 1, and 0.8 in image 2. Before filter implantation, the PSNR values range from 38.4532 to 36.2365, with a noise ratio between 0.3 and 0.5, but when the median filter values are used, the result is not as much modified, and the quality of the images remains similar, as the visual effects are much uncontaminated. The value of SSIM before filtration was between 5.3476 para-4 to 8.4536 para-4 and the noise ratio increased. The results showed that there was a minor difference between values, from 6.5287 para-4 to 5.234 para-3, immediately following the filtration results. The MSE values indicate the same variance. We calculate the same parameter flanked by the MRI image and the cross rim detector in order to interpret the hybrid edge detector output. This time, the PSNR values for salt-and-pepper noise range vary from 36.345 to 34.213. Compared to the original image, the principles of MSE and SSIM display similar quantity of variance. It clearly indicates that the developed technique is well known to retain image excellence.

6   Conclusion

In this research paper, we developed a method toward establishing a simple as well as suitable border in MRI images around the tumor area. Multi-phase medical image processing that involves filtration, edge detection, and segmentation is a way to achieve this form of results. The modern methodology of research is divided into three key processing phases. One of them is MRI image enrichment, which is carried out with medium filters in MRI images due to their salty and intersperse noise. Other filters cannot be used due to their blurring effect which removes the sharp distinctions of the edges and boundaries; the next step involves the detection of the cross-border to emphasize the limits of the brain tumor. In order to get acceptable consequences from border removal, we have used a hybrid edging operator that merges the most common edge operators. The K-means clustering is implemented in these areas. In combination with the results from dam design and the K-means clustering, hybrid technology improves the boundary of all structural cells in the MRI image. In conclusion, the final algorithm has been used to locate exactly the limit of the tumor area present in the MRI image specified. The reliability of the results was calculated after the usage of the technique suggested in some of the main parameters such as PSNR, MSE, and SSIM. In order to evaluate the precision of effects, these parameter values are measured at every step.

References

1. I. Mehidi, D.E.C. Belkhiat, D. Jabri, An improved clustering method based on K-means algorithm for MRI brain tumor segmentation, in *2019 6th International Conference on Image and Signal Processing and their Applications (ISPA)*, (IEEE, 2019), pp. 1–6
2. M. Faris, T. Javid, K. Fatima, M. Azhar, R. Kamran, Detection of tumor region in MR image through fusion of Dam construction and K-mean clustering algorithms, in *2019 2nd International Conference on Computing, Mathematics and Engineering Technologies (iCoMET)*, (IEEE, 2019), pp. 1–16
3. N. Mathur, P. Dadheech, M.K. Gupta, The K-means clustering based fuzzy edge detection technique on MRI images, in *2015 Fifth International Conference on Advances in Computing and Communications (ICACC)*, (IEEE, 2015), pp. 330–333
4. J. Selvakumar, A. Lakshmi, T. Arivoli, Brain tumor segmentation and its area calculation in brain MR images using K-mean clustering and Fuzzy C-mean algorithm, in *IEEE-International Conference on Advances in Engineering, Science and Management (ICAESM-2012)*, (IEEE, 2012), pp. 186–190
5. H. Parmar, B. Talati, Brain tumor segmentation using clustering approach, in *2019 Innovations in Power and Advanced Computing Technologies (i-PACT)*, vol. 1, (IEEE, 2019), pp. 1–6
6. S. Kumar, J.N. Singh, N. Kumar, An amalgam method efficient for finding of cancer gene using CSC from micro array data. Int. J. Emerg. Technol. **11**(3), 207–211 (2020)
7. S. Kumar, A. Negi, J.N. Singh, A. Gaurav, Brain tumor segmentation and classification using MRI images via fully convolution neural networks, in *2018 International Conference on Advances in Computing, Communication Control and Networking (ICACCCN)*, (IEEE, 2018), pp. 1178–1181

8. V.K. Lakshmi, C.A. Feroz, J.A.J. Merlin, Automated detection and segmentation of brain tumor using genetic algorithm, in *2018 International Conference on Smart Systems and Inventive Technology (ICSSIT)*, (IEEE, 2018), pp. 583–589

9. S. Kumar, A. Negi, J.N. Singh, H. Verma, A deep learning for brain tumor MRI images semantic segmentation using FCN, in *2018 4th International Conference on Computing Communication and Automation (ICCCA)*, (IEEE, 2018), pp. 1–4

10. S. Kumar, A. Negi, J.N. Singh, Semantic segmentation using deep learning for brain tumor MRI via fully convolution neural networks, in *Information and Communication Technology for Intelligent Systems*, (Springer, Singapore, 2019), pp. 11–19

11. R.M. Sumir, S. Mishra, N. Shastry, Segmentation of brain tumor from MRI images using fast marching method, in *2019 IEEE International Conference on Electrical, Computer and Communication Technologies (ICECCT)*, (IEEE, 2019), pp. 1–5

12. M. Thilagam, K. Arunesh, A. Rajeshkanna, Analysis of brain MRI images for tumor segmentation using Fuzzy C means algorithm, in *2020 International Conference on Electronics and Sustainable Communication Systems (ICESC)*, (IEEE, 2020), pp. 183–186

13. S. Pang, D. Anan, M.A. Orgun, Y. Zhezhou, A novel fused convolutional neural network for biomedical image classification. Med. Biol. Eng. Comput. **57**(1), 107–121 (2019)

14. A. Sellami, H. Hwang, A robust deep convolutional neural network with batch-weighted loss for heartbeat classification. Expert Syst. Appl. **122**, 75–84 (2019)

15. A. Wadhwa, A. Bhardwaj, V.S. Verma, A review on brain tumor segmentation of MRI images. Magn. Reson. Imaging **61**, 247–259 (2019)

16. H. Khan, P.M. Shah, M.A. Shah, S. ul Islam, J.J.P.C. Rodrigues, Cascading handcrafted features and Convolutional Neural Network for IoT-enabled brain tumor segmentation. Comput. Commun. **153**, 196–207 (2020)

Estimation of Human Posture Using Convolutional Neural Network Using Web Architecture

Dhruv Kumar, Abhay Kumar, M. Arvindhan, Ravi Sharma, Nalliyanna Goundar Veerappan Kousik, and S. Anbuchelian

1 Introduction

1.1 Pose Estimation

With web-based exercise videos and well-known fitness monitoring applications like Samsung Wellness, more people are individually pursuing these workouts. A significant number of individuals have embraced working independently indoors because of busy lifestyle and economic problems. Although easy, it can lead to significant injuries in the long term if the exercise is not done correctly. There is no mechanism at present to tell a customer how precisely he/she will obey a specific training. We propose a framework that can be used to evaluate and diagnose errors for any camera-based computer capable of streaming user workouts. The system is versatile enough to accommodate sync users and comparative images, together with any camera orientation objects (tilt, user size, user frame position). To solve the dilemma, our architecture uses deep learning and computer vision techniques.

The successful and precise identification of the components of the human body (pose) is a critical step in our process. In real-time multi-person 2D pose estimation using part affinity fields, we estimate human pose. To obtain a fixed size vector representation for a given image, the network uses the first ten layers of VGG19 net, which is followed by two multi-step branches of convolutional neural networks (CNN), where the first branch is used to predict confidence maps as defined by dots for body part (point) locations. The spine, elbows, wrists, shoulders, knees, hips,

D. Kumar (✉) · A. Kumar · M. Arvindhan · R. Sharma · N. G. V. Kousik
Golgotias University, Greater Noida, Uttar Pradesh, India
e-mail: dhruv.kumar@galgotiasuniversity.edu.in; abhaykumar@galgotiasuniversity.edu.in;
m.arvindhan@galgotiasuniversity.edu.in; ravi.sharma@galgotiasuniversity.edu.in

S. Anbuchelian
Ramanajunam Computing Centre, Anna University, Chennai, Tamil Nadu, India

© Springer Nature Switzerland AG 2022
P. Johri et al. (eds.), *Trends and Advancements of Image Processing and its Applications*, EAI/Springer Innovations in Communication and Computing, https://doi.org/10.1007/978-3-030-75945-2_12

eyes, and ears are part of the body. Vector field maps are predicted by the second branch, which helps to deduce the relation between body parts obtained from the first branch using the matching bipartite graph [1]. The human body is represented as a set of limbs identified by the keypoints once we obtain the key body points. An optimization to remember is that when conducting an exercise, not all limbs will be used; limbs involved in any motion will be used for the study portion.

With that in mind, we suggest a benchmark method for workout poses wherein the person records an activity video of him, and provide a methodical assessment [2, 3].

1.2 Pose Estimation Metrics

The exercise is performed by the individual in the captured video. An exercise rep E includes a series of z_i frames, where i indicates the frame number.

$$E = \{\text{frameset} z_i\}$$

In addition, where the fitness video includes a significant number of frames, the number of frames is limited to just the frames appropriate to the comparison of the workout.

$$\text{Relevant} - E = \{z - i \text{ pictures so that } z(i+1) \text{ image } z_i > \mu\}$$

Here μ is the distance between both the exercises frame E such that the frames within the distance can be skipped if the distance is smaller than μ. μ is measured on the basis of the angle shifts in continuous frames between sections of the body [4].

Finally, the E_u user exercise is paired with the E_t teacher exercise to verify the deviations. The relation is rendered by using the time series data-alignment algorithm DTW for matching frame pairs. First, the following metrics are widely used to predict poses:

- PCP – Right piece percentage proportion of the correct components contained on a given image [5]
- PCK – Right keypoint percentage
- PDJ – Percentage of joints found

When less than half the body part is the difference between actual joint positions and the two predicted joint positions, the person's body should be detected accordingly. Furthermore, PCP offers the following information:

- Effective body part identification rate. The biggest drawback is that shorter portions of the body are often penalized because shorter body parts have lower thresholds.
- Increasing the PCP value and then optimizing the model.

- PCK reflects the percentage of right keypoints from a given posing picture. If the difference in between actual body joints and the projected body joint is within a defined threshold, the joints of the body shall be described correctly. The threshold could be:

  - PCKh@0.5: if the criterion is 50% of the bone attachment of the brain
  - PCKh@0.2: Here the interval of values from the expected < true joint (0.2 × torso diameter) is <

- Often 150 mm is taken as a threshold.
- Decreases the shortness of the body portion since shorter sections of the body have wider torsos and head bone connections.
- PCK for 2D and 3D implementations. The greater the PCK rating, the better the model.
- PDJ is the amount of joint observed. If there is a certain fraction of the torso diameter between the true joint and the expected body joint, the observed joint should be accurately detected.
- PDJ@0.2 = actual joint distance < 0.2 × torso diameter expected.

2   Literature Review

2.1   Background and Techniques

The following Table 1 contains the name, author names, and merits and demerits of the all the recent published papers in the field of pose estimation.

- The standard approach to calculating poses relies mainly on visual templates. The central concept is to portray the human body as a non-rigid set of "sections." The component may be named as a matching beauty prototype. By springs, two separate sections are spatially related. The components are further located at their respective location and orientation such that a structured prediction task can be done by the system by modeling joint. However, the key constraint of this method is that the image data are not used in constructing a model [17].
- You can derive joint knowledge from film and perspective frames using a kinetic sensor. Needs an image processing technique to be applied.
- Use of EMG electrodes and use of heart rate sensors, etc.
- Trajectory-based human behavior detection probabilistic approach for defining the main incidents (basketball).
- FINA09 diving set: The following can be used to approximate poses: time data in video-oriented gradient histogram (HOG) in images [14].
- Quality evaluation: Train method uses action data collection of quality ratings. After preparation, the sample and performance score for consistency is measured between zero and 100 (Olympic diving data collection with judges) [18, 19].

Table 1 Distinct research field of pose estimation

S. no.	Resource	Merits	Demerits
1	Evaluation of spinal posture using Microsoft Kinect [6, 7]	Can extract information from video and depth frames	Only the video and depth frames can be found
2	Trajectory-based human activity recognition from videos [8]	Sparse paths could deliver equivalent or higher outcomes than dense pathways	Trajectory methods are not flexible to implement
3	Pose estimation based on PnP algorithm for the racket of table tennis robot [9, 10]	A robot device imitating human gestures like stroke and playing human beings	Imitation models require demonstration data or some way to acquire a supervised signal of desired behavior
4	DeepPose: Human pose estimation via deep neural networks [1]	First major paper to apply deep learning concepts. CNN regression toward body joints in a holistic fashion	Regression to XY coordinates can be difficult and adds learning complexity. Hence, it poorly performs in certain regions
5	Efficient object localization using convolutional networks [11]	Heatmap estimates the possibility of joint in each pixel. Heatmaps provide better results as compared to direct joint regression	Heatmap methods lack structure modeling
6	Convolutional pose machines [12, 13]	Completely differentiable and multi-stage architecture. Provides sequential framework for learning human pose	The lack of gradients for deep multi-stage networks is a significant complaint
7	Human pose estimation with iterative error feedback [14]	Self-correction model's original solution by iterative error eventually improves feedback	Limited to 2D pose estimation. Complex to implement
8	Stacked hourglass networks for human pose estimation [15, 16]	Captures global and local information completely and uses it to make predictions. Works for heavy occlusion and multiple people in close proximity	Some difficult pose cases not handled properly by the system. Complex to implement. Intermediate heatmap supervision is required

- The most adopted building block to approximate poses in recent years is the convolutional neural network (CNN), replacing handmade attributes and visual simulation. DeepPose by Toshev et al. used a deep learning technology to approximate poses by formulating poses as a Convolutionary dilemma of neural network-based regression into the joints of the human body. It had 7 AlexNet backend layers and a final layer, generating 2k joint positions [20].

$$\left(p_i, q_i\right) \times 2 \text{ for } i \in \{1, 2, 3, 4 \ldots n\} \text{ where } n = \text{number of body joints} \tag{1}$$

The model (Eq. 1) is trained with L2 loss on FLIC and LSP datasets for regression. It helps to optimize forecasts through cascading regressions. The regression to XY coordinates, however, is difficult and complicated. In few countries, it thus performs poorly [21].

3 Modules of the Proposed System

3.1 Human Pose Extraction

The extraction of poses is one of the main computer vision research areas. It can be human to distinguish objects within an image. There are separate methods for modeling and measuring objects in 2D and 3D. OpenPose, DeepCut, AlphaPose, Mask RCNN, and many other implementations for human pose estimation exist.

Place retrieval is among the most significant active research fields. It may vary from the human role to the recognition of various objects in an image. Different models exist to estimate the 2D and 3D extraction and object estimation. OpenPose, DeepCut, AlphaPose, Mask RCNN, and several other human pose estimates are available [12, 20, 22, 23].

OpenPose [14] is used to extract poses. It is one of the most common methods for estimating multiple poses. It uses the bottom-up method for the estimation of poses and starts with the use of the first several layers to remove image attributes. The functions are then moved to the CNN layer's parallel branches. One branch forecasts a total of 18 confidence maps such that a certain portion of the human skeleton is depicted on each map. The other branch forecasts 38 PAFs (part affinity fields). Such areas are being used to indicate the extent to which the pieces are related [24] (Fig. 1).

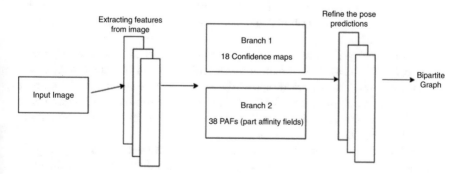

Fig. 1 OpenPose flowchart

Fig. 2 Bipartite graph
demo on multi-people
image

The following steps are used to optimize each branch's forecasts for poses. Bipartite diagrams are created by part trust maps between pairs of pieces. The weaker ties in the two-part graph are omitted on the basis of PAF values. By repeating these steps, a skeleton of the human pose is created for every person in the picture.

A two-party human pose consists of nodes and ties. The node indicates the joints and connections of the body which are linked.

The example of a bipartite graph for human pose is shown in Fig. 2.

Convolutional Neural Network

CNN or ConvNet is also regarded as a groundbreaking neural network. It is among the deep learning algorithms. It is an input for an image. The picture may be a video frame or a snapshot clicked with the camera. This algorithm assigns trained weights and distortions to different artifacts in the image after obtaining the input. The algorithm will separate one image from another image using these learning weights and biases. The neural network is favored here over feed-forward neural networks. A picture is a pixel value matrix. The CNN can capture the time and space requirements found in an image effectively with suitable filter. In addition, the CNN architecture offers weight re-usability and minimized parameters involved in picture creation (Fig. 3).

Convolution layers are the key structure of the CNN model. A convolution means that a target filter is added to the input image and the effects are achieved after activation. If the same filter is used repeatedly on the input image, the input produces the resulting output that is called a feature map after activation across a number of layers. The function map shows the location and intensity of the characteristics in the input image [25].

The filter for the transforming 2D input image is obtained by multiplying the input image data array with the 2D weight array. Often filters are often called as a kernel. The replication to acquire the filter or kernel is the dot product, that is, the

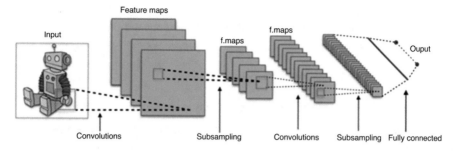

Fig. 3 CNN architecture

Fig. 4 CNN feature map

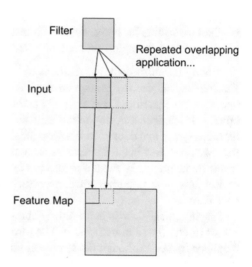

elementary multiplication of the filter and the filter portion of the input. It adds more and points to a single value. This technique is also known as a scalar product.

The same filter is applied on several input patches from left to left to right and top to bottom. If the function detects a function, the same filter may be used anywhere around the input image to recognize a certain feature. This is sometimes considered an invariance of translation (Fig. 4).

By testing the convolutionary network on unique data sets, the weights for these features are achieved. The convolutional neural network will connect to a given image from 32 to 512 filters in parallel and benefit from them. CNN layers can be extended not only to raw pixel values but to outputs from other layers as well. The piling of convolution layers results in a work-breakdown structure of the input signal [26]. If the filer is worked directly, low-level features such as lines are taken from raw pixel values. These lines are then moved to another layer of convolution to remove some lines the system can be changed dominated types. This method is persisted until the role is disabled. The abstraction of attributes advances to higher orders as the depth of the convolution network increases. In comparison, pooling layers are being used to reduce the spatial scale of features. This lowers the processing power needed and increases the efficiency of the convolutionary model.

Monitoring of Optical Flow

A video can contain several frames. This is why monitoring optical flow is used:

- Minimize the overall number of frames that are meaningless; compact the video.
- Balance the video if those frames are missing.
- Produce movement form.

It supposes that:

- Pixel intensities for two successive frames do not shift.
- The pixels in the opposing *image have identical rotation patterns.*

Among the most common methods for optical flow control is the L-K (Lucas-Kanade) technique. It operates on the basis of the above principles and sets the optimal flow equations for each of the pixels in the image with the least square fits criteria. It eradicates in the optical equation the ambiguity inferred. Furthermore, there are sparse and dense models for optical flow. Sparse optical flow offers vectors for edge pixel flow in the camera. Dense optical flow provides the flow vector for all pixels in the image. While the precision of dense optical flow is better than a small optical flow, the computing power necessary for dense optical flow is high, which is so lenient.

Sparse optical flow is achieved by choosing an edge or corner from the input vector (motion) and monitoring it. The forecast track is validated by inspection of neighbor frames. Although dense optical flow is the same as sparse optical flow, it is used for all the pixels and primarily for video segmentation and structure gathering from movement. OpenCV is the library for sparse or dense optical flow implementation [27]. Figure 5 displays the optical flow monitoring of a video vehicle using the OpenCV Lucas-Kanade system.

Fig. 5 Sparse optical flow tracking using OpenCV

Algorithm of Time Series Data Alignment

DTW is one of the most common algorithms for time series data alignment. These algorithms are used to compare the resemblance between two-time sequences. Both sequences can operate at various speeds. For instance,

- Comparison of the person's walking style from two recordings
- Comparison of speech styles of two people

DTW stands for time warping dynamics. The rules and limitations for matching two videos are as follows:

- Each first video frame must complement one or more second video frames or vice versa.
- The first video frame must fit in the second video's first frame. (More matches than this frame can be found.)
- The very last first video frame must match the last other video frame. (The matches should be more than this frame.)
- The first video's mapping of frames to frames from other videos must be monotonous or vice versa.
- The two-time sequences are originally separated into equivalent points.
- The Euclidean interval between the first sequence point and any point in the other sequence in time is measured. For reference, the distances are stored. This stage is also called time warp.
- Next, in the first video, travel to the second point, and then measure the distance to each point in the other time.
- The steps above are repeated before the first series endpoint meets.
- The above steps are repeated by using the second series as a reference.
- All minimal stored distances add up to have a graph of similarities between two periods (Fig. 6).

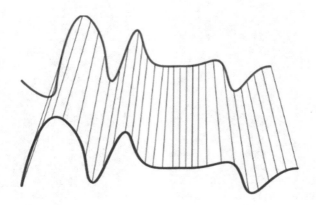

Fig. 6 DTW mapping

If a match follows the above criteria and has a low cost, the match is said to be ideal. The time warping algorithm produces a nonlinear alignment between two sequences, i.e., the optimal alignment possible between two sequences is achieved. It functions therefore as a strong indicator of similarity even though the two sequences are out of phase [28].

- DTW's complexity is $O(p \times q)$, where p and q are time series lengths.
- PrunedDTW, FastDTW, and SparseDTW are the fastest combinations of DTW.

Description of Framework

OpenPose by Cao et al. offers real-time 2D pose predictions by using component affinity fields. For a specific frame image, ten layers of VGG-19 net are used to provide a vector representation in a fixed size.

The first multi-step division describes the spine, shoulders, knee, legs, eyes, elbows, and ears, while, as seen in Fig. 7, the second multi-stage division describes the relation between body parts identified using bipartite diagrams. The exercise will require a variety of changes in the role [29, 30].

Table 2 indicates the variations of angles between the elements taken into consideration during the study of the exercise frames. The video of the trainer is first preprocessed using the OpenPose for each frame. In real-time situations, it is not desirable to perform CNNs on each patient video frame because the effects are latent. That is why optical flow monitoring is achieved such that visual feedback is generated more easily. Every nth frame (e.g., every fourth or eighth frame) is collected from the patient's video and then optical flow monitoring is used to collect body part keypoints for the intermediate frame. It offers the following benefits:

- Reduce errors in reverse and forward tracking estimations.
- Approximate keypoints in a missed context using neighboring frame information.

Fig. 7 Pose extraction

Table 2 Variations of angles between the elements

Body keypoint-1	Body keypoint-2	Body keypoint-3
Nose	Neck	Shoulder (right)
Nose	Neck	Shoulder (left)
Neck	Shoulder (right)	Elbow (right)
Neck	Shoulder (left)	Elbow (left)
Shoulder (right)	Elbow (right)	Wrist (right)
Shoulder (left)	Elbow (left)	Wrist (left)
Neck	Hip (right)	Knee (right)
Neck	Hip (left)	Knee (left)
Hip (right)	Knee (right)	Ankle (right)
Hip (left)	Knee (left)	Ankle (left)

The angles between different body parts is formed by Body keypoint-1, Body keypoint-2 and Body keypoint-3, such that Body-keypoint-2 is at the vertex

Next, DTW (dynamic time warping) is one of the most used algorithms for comparing two time series sequences. Two wave sequences aligned using the DTW algorithm are shown in Fig. 8. The applications of DTW include the comparison of walking styles of two people (different speeds), speech recognition from two speakers (different word rates), etc. [31].

In DTW, frames act as the similarity measure and help in distinguishing the differences between the two videos obtained from the trainer and patient. The metric (Eq. 2) for the distance between two frames is defined as S_{pq}.

$$S_{pq} = \sum_n \left| \text{body partangle}_{\text{patient}} - \text{body partangle}_{\text{trainer}} \right| \tag{2}$$

S_{pq} is the distance between pth frame in patient's video and qth frame in trainer's video and n is the set of body part angles used in the current exercise. The similarity between the patient's and trainer's video is calculated using DTW as shown in the Fig. 8. It is possible to have multiple frames from the first sequence mapped to a single frame in the second sequence (e.g., patient in a steady pose).

The plot includes frames of the specialist Y-axis video and frames of the patient X-axis video. The line plot should be straight in ideal shape, which implies that the exercise undertaken by the patient is perfect and there are no deviations. After the frame mapping, the next step is to carry out affine transformations to eliminate any potential problems of perception (image orientation, person to camera distance, patient and expert's separate body ratios). The expert frame is turned into patient points (or vice versa) and the skeleton overlaps on the expert frame. The least problem of squares is solved to produce a sophisticated transformation matrix. The transformation matrix T is seen as (Eq. 3)

$$TX = Y \tag{3}$$

Fig. 8 DTW algorithm
similarity graph

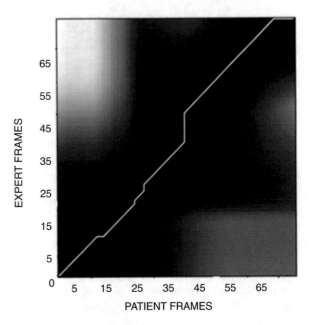

where X is the patient's body keypoint coordinates and Y is the expert's body key-point coordinates. The optimization is calculated as (Eq. 4)

$$Min = (TX\text{-}Y\text{-}2) \tag{4}$$

The flowchart in Fig. 9 shows the overview of the system for pose estimation and correction.

The patient uploads an ideal preparation video to the benchmark scheme for the preparation. The machine then collects and records the positions and angles of the joints using a CNN model. Later, the user uploads the exercise footage to the exercise benchmarking device at home. This video can be taken by tablet, laptop, or some other computer and the computer can be set in an angle. Each nth frame shall be extracted with coordinates and angles of joint keypoints from this film. The next step is to measure the intermediate frames and the time series data-alignment algorithm for the framemapping with optimum flow monitoring. The machine will now match the two videos using advanced transformations. Mistake occurs as a visual disparity between the joint keypoints derived from the footage of the patient and the specialist [31].

The plot in Fig. 10 demonstrates how the angle between expert and patient's forearm and upper arm varies frame by frame. The map can be drawn depending on which portion of the body is engaged more in exercise.

The device is based on the above plot to show the message "Hold your arms closer" or "not spread your arms too far." This allows the user to gain input from the device and change the workout accordingly [8, 32].

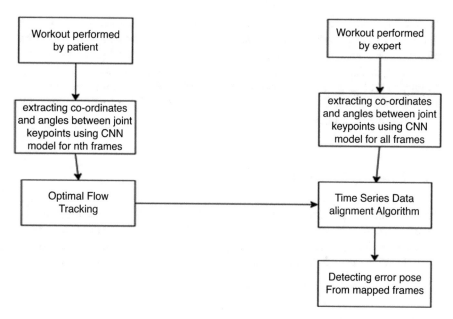

Fig. 9 System overview flowchart

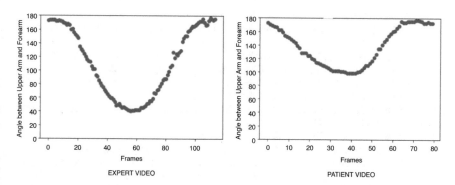

Fig. 10 Expert and patient plot comparison

4 Conclusion and Future Scope

4.1 Conclusion

The images were taken at a frame rate of 25 fps. The 1280 × 720 resolution laptop and camera is used for patient action monitoring. The efficiency of the method is assessed based on the number of frames that were larger than μ. The variance is computed on the basis of the maximal angles between the angles of the body component and the micro symbol shows the threshold on that angle. DTW algorithm improves the performance of the frame matching method. The comparison video of the specialist was played and patients were asked to conduct the same procedure representative as experts during the experiment. The device was tested by 4–5 volunteers of various physical. The volunteers were very happy with the results. This paper deals with a teaching benchmark framework to support patients with physiotherapy. It will allow people to correct themselves and get as far from their home sessions as possible. The system can detect the exercise deviation that is important in home physiotherapy sessions.

4.2   Scope of the Future

The abovementioned CNN model was trained on the COCO data collection. The device can now only sense motion in two dimensions but in future prediction can be expanded with depth perception technologies or external camera along three dimensions. In future, the CNN model can be optimized for quicker inference.

References

1. A. Toshev, C. Szegedy, DeepPose: Human pose estimation via deep neural networks, in *The IEEE Conference on Computer Vision and Pattern Recognition (CVPR)*, (IEEE, 2014), pp. 1653–1660
2. B. Xiao, H. Wu, Simple baseliners for human pose estimation and tracking, in *The European Conference on Computer Vision (ECCV)*, (Springer, 2018), pp. 466–481
3. Z.G. Cao, OpenPose: Realtime multi-person 2D pose estimation using part affinity fields. Comput. Vis. Pattern Recognit. arXiv **1812**, 08008 (2018)
4. Z.S. Cao, Realtime multi-person 2D pose estimation using part affinity fields. arXiv preprint arXiv:1611.08050 (2016)
5. A. Haque, B.-F. Peng, Towards viewpoint invariant 3D human pose estimation, in *Computer Vision – ECCV. Lecture Notes in Computer Science*, vol. 9905, (Springer, 2016)
6. E.L. Insafutdinov, Deepcut: A deeper, stronger, and faster multiperson pose estimation model, in *Computer Vision – ECCV. Lecture Notes in Computer Science*, vol. 9910, (Springer, 2016)
7. A. Castro et al., "Evaluation of spinal posture using Microsoft Kinect™: A preliminary case-study with 98 volunteers", Porto Biomedical Journal, vol. 2, no. 1, pp. 18–22, 2017. Available: https://doi.org/10.1016/j.pbj.2016.11.004.
8. B. Boufama, Trajectory-based human activity recognition from videos, in *International Conference on Advanced Technologies for Signal and Image Processing* -, no. 978-1-5386-0551-617, Accessed 22 May 2017, (2017), pp. 1–5
9. A. Jain, J. Tompson, MoDeep: A deep learning framework using motion features for human pose estimation, in *Computer Vision – ACCV. Lecture Notes in Computer Science*, vol. 9004, (Springer, 2014)
10. J. Wang, L. Shi, S. Song, J. Tian, X. Kang, Tetraploid production through zygotic chromosome doubling in Populus. **47**(2), article id 932 (2013). https://doi.org/10.14214/sf.932
11. J. Tompson, R. Goroshin, Efficient object localization using convolutional networks, in *The IEEE Conference on Computer Vision and Pattern Recognition (CVPR)*, (IEEE, 2015), pp. 648–656

12. S.H. Ren, Faster R-CNN: Towards real-time object detection with region proposal networks, in *Advances in Neural Information Processing Systems*, (2015), pp. 91–99
13. S.-E. Wei, V. Ramakrishna, Convolutional pose machines, in *The IEEE Conference on Computer Vision and Pattern Recognition (CVPR)*, (IEEE, 2016), pp. 4724–4732
14. J. Carreira, P. Agrawal, Human pose estimation with iterative error feedback, in *The IEEE Conference on Computer Vision and Pattern Recognition (CVPR)*, (IEEE, 2016), pp. 4733–4742
15. A. Newell, K. Yang, Stacked hourglass networks for human pose estimation, in *Computer Vision – ECCV. Lecture Notes in Computer Science*, vol. 9912, (Springer, 2016)
16. Y. Sun, Deep convolutional network cascade for facial point detection, in *The IEEE Conference on Computer Vision and Pattern Recognition (CVPR)*, (IEEE, 2013), pp. 3476–3483
17. J. Shotton, T. Sharp, Real-time human pose recognition in parts from single depth images. Commun. ACM **56**(1), 1–8 (2013)
18. A.B. Kanazawa, End-to-end recovery of human shape and pose, in *The IEEE Conference on Computer Vision and Pattern Recognition (CVPR)*, (IEEE, 2018)
19. F.A. Kondori, A direct method for 3D hand pose recovery, in *22nd International Conference on Pattern Recognition (ICPR)*, (IEEE, 2014), pp. 345–350
20. M.W. Oberweger, Hands deep in deep learning for hand pose estimation, in *Proceedings of 20th Computer Vision Winter Workshop (CVWW)*, (Graz University of Technology, 2015), pp. 21–30
21. C.A. Szegedy, Object detection via deep neural networks. *NIPS*, vol 26 (2013)
22. M.W. Oberweger, Training a feedback loop for hand pose estimation, in *Proceedings of the IEEE International Conference on Computer Vision*, (IEEE, 2015), pp. 3316–3324
23. L.E. Pishchulin, DeepCut: Joint subset partition and labeling for multi person pose estimation, in *The IEEE Conference on Computer Vision and Pattern Recognition (CVPR)*, (IEEE, 2016)
24. G.S. Tamas, Body part extraction and pose estimation method in rowing videos. J. Comput. Inf. Technol. **26**(1), 29–43 (2018)
25. G.W. Taylor, Pose-sensitive embedding by nonlinear NCA regression. *NIPS* (2010)
26. Y. Tian, Exploring the spatial hierarchy of mixture models for human pose estimation, in *Computer Vision – ECCV. Lecture Notes in Computer Science*, vol. 7576, (Springer, 2012)
27. J.S. Tompson, Real-time continuous pose recovery of human hands using convolutional networks. ACM Trans. Graph. **33**(5), 169 (2014)
28. A.A. Toshev, DeepPose: Human pose estimation via deep neural networks, in *The IEEE Conference on Computer Vision and Pattern Recognition (CVPR)*, (IEEE, 2014)., http://human-pose.mpi-inf.mpg.de
29. F.A. Wang, Beyond physical connections: Tree models in human pose estimation, in *The IEEE Conference on Computer Vision and Pattern Recognition (CVPR)*, (IEEE, 2013)
30. Y.A. Yang, Articulated pose estimation with flexible mixtures-of-parts, in *The IEEE Conference on Computer Vision and Pattern Recognition (CVPR)*, (IEEE, 2011)
31. M. Ye, Q. Zhang, A survey on human motion analysis from depth data, in *Time-of-Flight and Depth Imaging. Sensors, Algorithms, and Applications. Lecture Notes in Computer Science*, vol. 82, (Springer, Berlin, Heidelberg, 2013)
32. S. Yuan, G. Garcia-Hernando, Depth-based 3D hand pose estimation: From current achievements to future goals, in *The IEEE Conference on Computer Vision and Pattern Recognition (CVPR)*, (IEEE, 2018)

Histogram Distance Metric Learning to Diagnose Breast Cancer using Semantic Analysis and Natural Language Interpretation Methods

D. Gnana Jebadas, M. Sivaram, Arvindhan M, B. S. Vidhyasagar, and B. Bharathi Kannan

1 Introduction

Natural language processing (NLP) is a multidisciplinary field that uses computational methods:

- To investigate the properties of written human language and to model the mental properties underlying the accepting and creation of written communication (with a scientific focus);
- To develop new practical applications involving computer-based smart processing of written human language (with an engineering focus).
- *Natural Language Processing* (NLP) is a subset of artificial intelligence and linguistics committed to having computers be able to recognize statements or words written in human languages. NLP came into being in order to alleviate the work of computer users and to achieve the goal of communicating in natural language with the machine because not all users will be able to become familiar with the specific machine languages [1].

D. G. Jebadas
School of Electronics and Communication Engineering, Galgotias University,
Greater Noida, Uttar Pradesh, India

M. Sivaram
Research Center, Lebanese French University, Erbil, Iraq

A. M (✉) · B. B. Kannan
School of Computing Science and Engineering, Galgotias University,
Greater Noida, Uttar Pradesh, India
e-mail: saroarvindmster@gmail.com

B. S. Vidhyasagar
Department of Computer Science and Engineering, Amrita School of Engineering, Amrita
Vishwa Vidyapeetham, Chennai, Tamil Nadu, India

© Springer Nature Switzerland AG 2022
P. Johri et al. (eds.), *Trends and Advancements of Image Processing and its Applications*, EAI/Springer Innovations in Communication and Computing,
https://doi.org/10.1007/978-3-030-75945-2_13

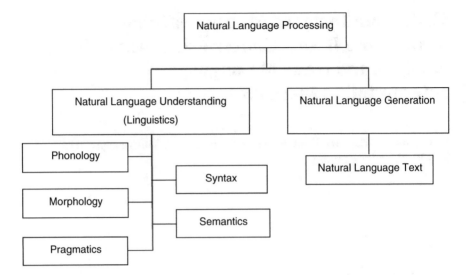

Fig. 1 NLP classification

- A language can be defined as a set of rules or a collection of symbols. Images remain pooled and used aimed at persuasive data or propagating the data. Symbols stay stamped in sudden practicalities. The linguistic communication process may generally be classified into two elements: understanding linguistic communication and generation of linguistic communication; out of these two elements evolve the tasks needed to understand and generate text-based communication (Fig. 1).

Linguistics comprises the language science that involves sound-related phonology, word forming morphology, syntax sentence formation, semantics grammar, and understanding related pragmatics.

One of the twentieth century linguists to introduce syntactic theories, Noah Chomsky, marked out a unique place for himself in the field of theoretical linguistics, which transfigured the world of syntax (Chomsky, 1965) [2]. Syntax can be generally classified into two levels: a higher level that incorporates speech recognition, and a lower level that corresponds to natural language. Some of the research topics in NLP are automatic summarization, co-reference resolution, discourse analysis, artificial intelligence, morphological segmentation, named entity recognition, optical character recognition, and part-of-speech tagging. A number of these have real-world applications, such as artificial intelligence, named entity recognition (NER), and optical character recognition. Automatic summarization generates an understandable summary of a set of text. Co-reference resolution refers to a word or expression that stipulates the word that refers to the same subject. Discourse analysis identifies the discourse structure of the related text. Machine translation automatically translates from one human language into another. Morphological segmentation refers to the separation of words into individual morphemes and to identification of the class of morphemes. Named entity recognition describes a text stream, determining which items in the text are proper names.

Challenges in restorative medicine involve familiarity with curative illustration statistics. The supply learning mechanism in medical interpretation is rather large. This is a priority for seeking the action of assortment and acceptance systems to aid medicine roborant experts in diagnosing diseases. The intent of intuitive Pidgin processing's is to fit together and send an algorithm or criterion metrics disciplines.

It has been used in bilingual incident discovery. Rangayyan et al. [3] intended to make a new modular system for cross-lingual dealings for Italian, Dutch, and English texts using dissimilar pipelines aimed at dissimilar languages. This scheme introduces a modular array for natural language processing (NLP) utensils.

These pipelines combine modules used at basic natural language processing as additional difficult tasks similar to cross-lingual termed entity linking, grammatical type labeling, then time normalization. Therefore, the cross-lingual agenda favors the understanding of measures, contributors, localities, and period [4]. Output of these separable pipelines is intended to be used as input for a scheme that acquires occurring central data graphs. All of the modules perform corresponding to UNIX operating system pipes: each of them yields commonplace input, to try and generate a few explanations, and turn out commonplace output that successively is the input on behalf of the consecutive module pipeline area unit designed with a known central design in order that modules are custom-made and replaced. Moreover, standard design permits for various configurations and for dynamic distribution.

Multi-hop reading comprehension (RC) demands definition and a combination of data exceeding several substance declaration objects. Throughout this initiative, we lean to dispute that big multi-hop RC datasets would be demanding to assemble. Multi-hop logic could also separate all the inquiries and even offer an indication of solution; often highly integrative inquiries can be answered with one hop whether they want comprehensive unit sorts or the actualities needed to respond to this region unit shortage [5, 6].

2 NLP Levels

Natural language processing has many "levels of language" that work together to obtain the human language technology text by attaining content with surface realization phases to retrieve information (Fig. 2).

2.1 Phonology

Phonology is the tagmemics connection that leads toward functional punctilious change. Phonology comes from Ancient Greek, where *phono* means *sound* or *voice*, and *-logy* is Greek and means *word, speech,* or *discourse.* In 1993, Nikolai

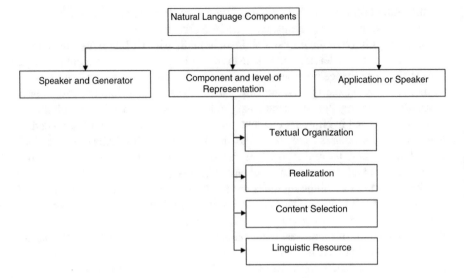

Fig. 2 NPL levels

Trubetzkoy assumed that well known wind phonology was "dissect of recommendable suitability for burr practices" [7].

2.2 Morphology

Morphemes are the minimum units of meaning of the dissimilar portions of speech. Morphology deciphers word nature through morphemes. An example of morpheme is that the word *pre-cancellation* can be morphologically broken down into three diverse morphemes: prefix *pre*, root *cancella*, and suffix *ation* [8].

2.3 Lexical

Lexical is similar to the human language system of vocabulary that includes distinct words. Miscellaneous sorts of procedures contribute to word-level understanding with the most important of those being a part-of-speech label to each word. Throughout this practice, words that can perform as more than one part-of-speech are chosen as the principal plausible part-of speech depending on the circumstance through which it happens. At the lexical level, linguistics depictions may have more than one meaning. In a human language technology system, the meaning of those characters are defined in line with the context of the other words.

2.4 Syntactic

The syntactic level, also called parsing, inspects the words to uncover the grammatical structure of the sentence. A synchronic linguistics computer program is necessary through this level. The output of the level can give a picture of the sentence absent its structural dependency. Several phrases might still be ambiguous to a computer program [9]. Not every information processing solicitation necessitates a complete analysis of sentences, thus deconstructing of phrase parts and combining examination at present does not encumber the request that phrase and grammatical construction dependencies are tolerable. Syntax conveys meaning in the majority of languages because of word order and supplying association. For example, the two sentences: "The cat hunted person the mouse." and "The mouse hunted person the cat." vary solely in stipulations for syntax; however, they express reasonably entirely dissimilar significance.

2.5 Semantic

In linguistics the public might presume the incorrect word meaning, and it is often the case that the entire sentence imparts the meaning. The linguistics process decides the feasible significances of a sentence according to word-level meanings in the sentence. These levels of process will integrate this linguistics clarification of words by various meanings into a syntactical explanation of words that may occur as several parts-of-speech at the syntactical level. For instance, surrounded by different connotations, "file" as a noun describes a folder that holds papers; "file" can also mean a tool to smooth one's fingernails; or a row of people in an exceptionally long queue). The linguistics level examines words to determine their possible meaning; however, the environment of the sentence helps determine the meaning. In the linguistics environment almost every word has more than one meaning; however, we are able to mark the acceptable one by viewing the remainder of the sentence [10].

3 Natural Language Generation (NLG)

Natural language generation (NLG) is a method for manufacturing meaningful phrases, paragraphs, and sentences for natural language. It is a type of language process and occurs in four segments [11]: plan the objectives coming up by appraising the case and obtainable expansive sources and understanding the content of plans (Fig. 3). Its contradictory output can be considerate.

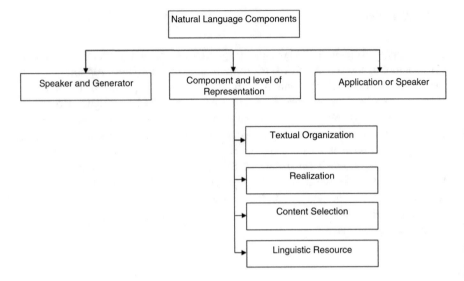

Fig. 3 NLG Components

3.1 Speaker and Generator

In order to create a message, we require a speaker or a solicitation with a generator or system that makes the purpose of the application applicable to the situation in a fluent sentence.

3.2 Representation Elements and Rates

The language development process involves the following interwoven activities. Selection of content: Data must be chosen and incorporated in the place. Depending on how this material is parsed into components of accountability, parts of the components might need to be detached, while a few others may need to be further processed [12].

3.3 Theological Organization

Knowledge must be arranged textually as shown by grammar, ordered manually and in terms of linguistic relationships such as changes. Semantic resources: linguistic services must be chosen to facilitate the comprehension of the knowledge. Eventually, these tools may result in the use of specific words, idioms, grammatical frameworks, etc.

3.4 Program or Speaker

This is just to hold the situation template. Here the speaker starts the process and does not engage in the exchange of languages. It stores the history, frameworks the potentially useful data, and executes a depiction of what it probably knows. Most of these shape the state of affairs; while choosing the sub-set of proposals, the presenter has only the criterion that the speaker should be aware of the issue [13].

4 Medicine

NLP is also useful in the medical field. The string of words project-medical language processor (PMLP) alone has a massive scale from IP within the field of medicine. The LSP-MLP assists sanctionative physicians to pull out and recapitulate info of any signs or indications, drug dose and reply knowledge with the endeavor to realize the potential aspect of some medication [14]. The National Library of drugs is growing the authority system. It is expected to operate as an info extraction tool for medicine information bases, notably telephone system abstract.

The lexicon has been developed using Medical Subject Headings (MeSH), Dorland's exemplified Medical Dictionary, and English General Dictionaries. The Hospital Cantonal de Geneve's Center d'Informatique Hospitaliere is working on an online archiving system using NLP. Patient records were archived in the first phase. Later on, the LSP-MLP was adapted for French, and finally, a proper NLP system called RECIT was developed using the location processing method. It had the job of introducing a comprehensive and multilingual program capable of analyzing and understanding clinical phrases and maintaining text-free information in an independent knowledge language [15].

New York University of Columbia has an urbanized NLP system called MEDLEE, which recognizes medicinal data in descriptive documents and converts textual information into organized representation.

5 Breast Cancer

Breast cancer can be diagnosed under different levels of development and different growth rates. Breast cancer is difficult to evaluate accurately as it is the world's most dangerous illnesses, and it can differ from one person to another [16]. To cure cancer, surgery and therapy techniques or a combination of the two are required depending on the grade, size, and stage of the cancer. Carcinoma is cancer that takes place in the lining of the organ (e.g., breast). Similar to other cancers, carcinoma is titled whether ductal or lobular. In situ in carcinoma applies to an initial stage of cancer, consigned to the layer of cells where it began. The cells are located in ducts in the class in situ, which is named in situ ductal carcinoma [17].

5.1   Categories of Breast Cancer

Breast cancer is stratified 1, 2, or 3. As per severity of growth, the third grade of neoplasm grows abnormally, and formation of neoplasm grows slowly. In DCIS, the three levels are termed as low, medium, and high severity. Throughout the first stages of cancer, we have a tendency to not be aware of it; however, later it is recognized through the abnormal changes within the size of the breast. Some hospitals would possibly use the neoplasm, node, and metastases (TNM) methodology [18]. It is an alternate way to represent the cancer severity level. During this methodology the dimensions of a neoplasm, the quantity of affected lymph gland, and severity of spread over all alternative components is set, and with the assistance of numbers and letters the stages are drawn. The description of a TNM process is typically defined as private and public.

If it is known as primary breast cancer within the breast, and if it is found beside the central region of the breast, it is categorized as supplementary breast cancer. The key tumor can be supported by lymph nodes as it takes the infected tissue to other parts of the human body. On the other hand, the same work is also performed by the bloodstream. Metastatic cancer is the common name of secondary breast cancer [19, 20] (Fig. 4 and Table 1).

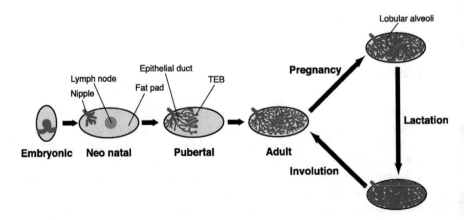

Fig. 4 Development of cancer cells

Table 1 Describing the types of mammography

Type of cancer	Severity	Characteristics	Affected age period
Ductal carcinoma in situ (DCIS)	Non-invasive	Affected cells spreads over ducts and other parts	Affects women who are over 50 years of age
Invasive ductal carcinoma	Invasive	Spreads over lymph nodes	Affects late 50s women
Lobular carcinoma in situ (LCIS)	Slightly invasive	Spreads to other breasts	Late 40s
Invasive lobular carcinoma	Invasive	Number of multicentricity in both the breasts	Late 40s
Inflammatory	Risky	Block in skin lymph	Late 50s
Tubular carcinoma	Non-invasive	Auxiliary lymph node affected	Late 50s
Medullary carcinoma	Slightly invasive	Infiltration in lymphocyte	Late 50s

5.2 Development of Cancer

6 Mammography Screening

It is a low-dose X-ray screening method when women do not have symptoms. Limited screening and screening will not detect all types of breast cancers, and an increased view is needed for women who have had breast implantation.

6.1 Mammography Diagnosis

This is a certain kind of technique for X-ray analysis. Necessary when an unusual alteration in one or both breasts occurs after a surveillance mammogram has detected any unusual findings. Side effects may include changes in mass, size and shape, bleeding of the nipple, skin density change, etc. This will help make sure that the unusual modifications belong to a class of benignity [21].

6.2 Mammography Online

It is possible to store electronic mammograms effectively on the PC. High-quality pictures can be obtained, and it allows a closer look if extended. Exposure is very small compared to video movie mammography. More expensive compared to all

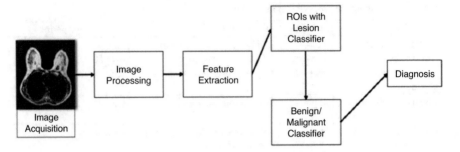

Fig. 5 Differentiating the image on a ROI basis. (**a**) Database 1 preparation of the extracted features. (**b**) Preparation of base 2 information from the functionality collected

other testing techniques. Regardless of a patient's age or pregnancy status, if suspicious things are identified, regular mammography is recommended.

6.3   Screen Film Mammography

IT is the most common type of breast exam treatment. The quality of the collected images is in full color shown on a film sheet. Quality is close to digital mammography. IT is specifically recommended to women over 50 years of age, still facing menstruation and unloading in the nipple, they have to go for either testing or digital mammography.

The identification application's general design is suggested with grouping. A filter pool first breaks down the selected server images into site images. Using hybrid DWT methods, site-images are broken down into four regions. The sub-images are collected using a hessian matrix and analyzed through the region of interest (ROI). The Bayes algorithm is used to differentiate a microcalcification cluster between abnormal ROIs. Therefore, we need to select ROIs at a slower period, if one of the ROIs will be at the middle of a microcalcification group [22, 23] (Fig. 5).

7   Conclusion

The zone of microcalcification can be identified and the existence of other microcalcification clusters is necessary to explain the diagnosis. The presence of microcalcification on the outcome image and the previous one can be identified by visual examination of the mammogram. Concentrated microcalcification could not be identified in a few samples, due to the test picture shown on the identification of microcalcification. Owing to an inability in the monitoring process, where only 1 image of all 30 test images failed, the simulation's usefulness can be reduced. The method does not require complete organic matter and rebuilding; it can be done easily. Complicated problems are enhancing the optimization and differentiation algorithms.

References

1. S. Grossberg, Competitive learning: From interactive activation to adaptive resonance. Cognit. Sci. **11**(1), 23–63 (1987)
2. G.A. Carpenter, S. Grossberg, ART 2: Self-organization of stable category recognition codes for analog input patterns. Appl. Opt. **26**(23), 4919 (1987)
3. R.M. Rangayyan, F.J. Ayres, J.E. Leo Desautels, A review of computer-aided diagnosis of breast cancer: toward the detection of subtle signs. J. Franklin Inst. **344**(3–4), 312–348 (2007)
4. G.A. Carpenter, S. Grossberg, D.B. Rosen, ART 2-A: An adaptive resonance algorithm for rapid category learning and recognition. Neural Networks **4**(4), 493–504 (1991)
5. G.A. Carpenter, S. Grossberg, ART 3: Hierarchical search using chemical transmitters in self-organizing pattern recognition architectures. Neural Networks **3**(2), 129–152 (1990)
6. http://gemi.mpl.ird.fr/PDF/Lek.EM.1999.pdf
7. http://www.ai-junkie.com/ann/som/som2.html
8. http://www.learnartificialneuralnetworks.com
9. http://en.wikipedia.org/wiki/Adaptive_resonance_
10. R. Alonso-Calvo et al., A semantic interoperability approach to support integration of gene expression and clinical data in breast cancer. Comput. Biol. Med. 87, 179–186 (2017). Available: https://doi.org/10.1016/j.compbiomed.2017.06.005
11. Natural Language Processing. Natural Language Processing. RSS. N.P., 2017.
12. Srihari S. Machine learning: generative and discriminative models 2010. http://www.cedar.buffalo.edu/wsrihari/CSE574/Discriminative-Generative.pdf. Accessed 31 May 2011
13. Elkan C. Log-linear models and conditional random fields 2008. http://cseweb.ucsd.edu/welkan/250B/cikmtutorial.pdf. Accessed 28 Jun 2011. 62. Hearst MA, Dumais ST, Osman E, et al. Support vector machines
14. D. Jurafsky, J.H. Martin, *Speech and Language Processing*, 2nd edn. (Prentice-Hall, Englewood Cliffs, 2008)
15. E.L.L. Sonnhammer, S.R. Eddy, E. Birney, et al., Pfam: Multiple sequence alignments and HMM-profiles of protein domains. Nucleic Acids Res. **26**, 320 (1998)
16. E.L. Sonnhammer, S.R. Eddy, E. Birney, A. Bateman, R. Durbin, Pfam: Multiple sequence alignments and HMM-profiles of protein domains. Nucleic Acids Res. **26**(1), 320–322 (1998)
17. Using Natural Language Processing and Network Analysis to Develop a Conceptual Framework for Medication Therapy Management Research. AMIA ... Annual Symposium proceedings. AMIA Symposium. U.S. National Library of Medicine, (2017)
18. W. Ogallo, A.S. Kanter, Using natural language processing and network analysis to develop a conceptual framework for medication therapy management research (2017). Retrieved April 10,2017, from https://www.ncbi.nlm.nih.gov/pubmed/28269895?dopt=Abstract
19. A. Ochoa, Meet the pilot: smart earpiece language translator (2016). Retrieved April 10, 2017, from https://www.indiegogo.com/projects/meet-the-pilot-smart- earpiece-language-translator-headphones-travel
20. S. Feldman, NLP meets the jabberwocky: natural language processing in information. Retrieval. Online-Weston then Wilton. **23**, 62–73 (1999).
21. E.D. Liddy, *Natural language processing* (Algorithmics Press, New York, 2001)
22. J. Berant, A. Chou, R. Frostig, P. Liang, Semantic parsing on freebase from question-answer pairs, in *EMNLP*, (2013)
23. N. Chomsky, *Aspects of the Theory of Syntax* (MIT Press, Cambridge, 1965)

Human Skin Color Detection Technique Using Different Color Models

Ruqaiya Khanam ⓘ**, Prashant Johri, and Mario José Diván**

1 Introduction

Humans are able to perceive a lot of color and many gray shades. Color image processing is based on two primary factors such as "pseudo- and full" color image processing. In "full-color" image processing, the true pictures are captured using full-color sensors like TV camera or scanner, whereas in pseudo-color image processing, it assigns the color to a range of monochrome intensities. Pseudo-colors are also known as indexed or false colors [1–4]. The light is an electromagnetic spectrum which is shown in Fig. 1.

Electromagnetic spectrum has a narrow band of frequencies for visible light. Human can perceive a formation of distinct wavelength spectrum. In the case of light, it means intensity is the only attribute of it. Gray level indicates a degree of intensity; the range starts from black to gray (with different shades of gray) and ended up to pure white. The chromatic light range starts from 400 to 700 nm as can be seen from Fig. 1 [4].

The quality of any particular color wavelength of light source can be defined using three basic quantities like "radiance," "luminance," and "brightness." Human eyes are able to perceive the primary colors which are red (65%), green (33%), and blue (2%) (Fig. 2).

The red colors' wavelength is 700 nm; similarly green and blue colors' wavelengths are 546.1 and 435.8 nm, respectively. Secondary colors are being produced

R. Khanam (✉)
Sharda University, Greater Noida, Uttar Pradesh, India

P. Johri
Galgotias University, Greater Noida, Uttar Pradesh, India

M. J. Diván
National University of La Pampa, Santa Rosa, Argentina
e-mail: mjdivan@eco.unlpam.edu.ar

© Springer Nature Switzerland AG 2022
P. Johri et al. (eds.), *Trends and Advancements of Image Processing and its Applications*, EAI/Springer Innovations in Communication and Computing,
https://doi.org/10.1007/978-3-030-75945-2_14

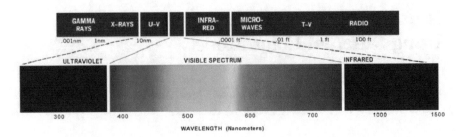

Fig. 1 Visible range of electromagnetic spectrum in wavelengths (nanometers). (Courtesy of General Electric Co)

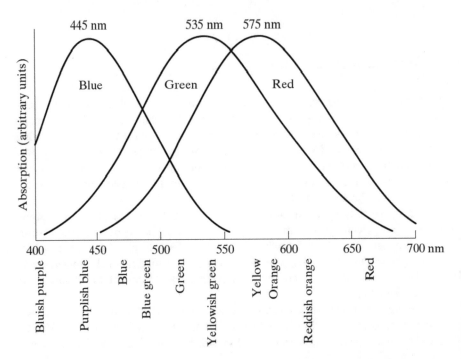

Fig. 2 Light absorption of the cone of red, green, and blue of the human eye (in terms of wavelengths) [1]

by primary colors but not in 100% (all color). The pigment or colorants can be defined as primary colors. Strong skin color detection starts to develop an efficient mathematical model using mathematical concepts to illustrate the information of skin color which is in the form of three (3) or four (4) different color components. Particularly these different color spaces (models) are RGB, HSV, YCbCr, and CMYK [1, 5] are widely used in the area of computer graphics, image processing, television telecasting, face recognition [5], gesture analysis, surveillance [6], skin disease [7–9], and computer vision. Skin recognition from an image is a main task to classify skin color means to identify skin and non-skin colors [7, 10–16]. Many other factors like surrounding lights show effect on its results. According to the recent research analysis, it reveals that the range of hues is restricted in human skin color which is not that deeply saturated. As a result, the formation of skin color is a combination of some specific colors like blood (in red color) and melanin (mixture of yellow color and brown color) [17–23]. In this research work, various color models would have been used in the skin detection process.

Our objective in this paper is to demonstrate the detection technique on different color models so that we could achieve better result for the required application. The paper is organized in different sections. Section 2 reveals many different color spaces for the purpose of detection process and basically it is an extension of the Introduction part. Section 3 covers all existing work on skin detection which is entirely related to this specific area. In Sect. 4 a new algorithm to calculate and evaluate its parameters is proposed. Section 5 reveals comparative results based on the various color models. Section 6 explains all results and discussion from different aspects. The conclusion is drawn in Sect. 7.

2 Color Models

The mathematical model of color space represents the information of color in the combination of three or four different color components [24]. However, the color of skin can be identified with the help of different color spaces. Four color patterns are discussed briefly.

2.1 Red, Green, and Blue (RGB) Model

RGB (red, green, and blue) and normalized RGB models might be used for color detection. The calculated value of filtered samples of each and every pixel is from red (R) and green (G) color in the form of normalized values. RGB color space or model has been used extensively. Furthermore, it is usually the primary color models for collecting and representing the digital images. Taking linear or nonlinear transformation of RGB space, we can get some other color space [25–28]. Skin color detection based on RGB color space is explained in [29–31]. Normalized RGB means normalized red, green, and blue. The way of representation can be easily attained from the values of RGB using a simple normalization procedure [30–33]. Therefore, it is showing its exceptional property which is most suitable for rough surfaces and is also ignoring an intense light. Due to its stable characteristics, certain assumptions of normalized RGB include its change in surface orientation which is relative to the light source [24, 33].

All the RGB values are falling in the range of zero and one. With the digital representation, for a fixed length of bits each color element. The total number of bits is

Fig. 3 RGB color
model [1]

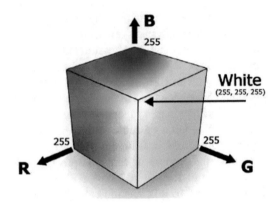

called color depth, or pixel depth. For example, in a 24-bit RGB color, each color
has 8 bits. The 8-bit binary number r represents the value of $r/256$ in [0, 1].

Using these three filters, a color image can be acquired which is more sensitive
to primary colors. If we want to capture any color using a monochrome camera
which is well equipped with any filter, the result will be an image in monochrome
but its intensity is proportionally based on the filters' response.

$$r = \frac{R}{R+G+B} \tag{1}$$

$$g = \frac{G}{R+G+B} \tag{2}$$

$$b = \frac{B}{R+G+B} \tag{3}$$

$$r + g + b = 1 \tag{4}$$

Expressions (1), (2), (3), and (4) represent the transformation from RGB to normalized RGB. Component "b" does not have any important information, although
this can be discarded when memory constraint is considered. Figure 3, RGB color
model shows the RGB color cube with their axis, and Fig. 4 demonstrates cross-
sectional plane by giving the inputs through color monitor and generating three
hidden surface planes which are shown in Fig. 5.

2.2 Hue, Saturation, and Value (HSV) Model

Another type of color model is based on hue like HSI, HSV, and HSL. This type of
approach discriminates the information of color and intensity under the condition of
non-uniform illumination condition. The process of conversion of color model such
as from RGB to HSI or HSV is time consuming and expansive as well. Due to

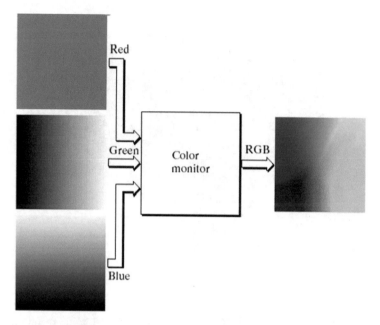

Fig. 4 Generation of RGB color model

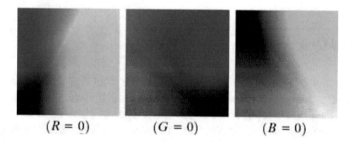

$(R = 0)$　　　　　$(G = 0)$　　　　　$(B = 0)$

Fig. 5 Hidden surface plane in the color cube obtained from primary colors

abrupt changes in color information such as hue and saturation, the pixel intensities whether they are small or large are not considered. HSV color model is illustrated Fig. 6 with its "hue, saturation, and value."

$$H = \arccos \frac{\left(2\text{Red} - \text{Green} - \text{Blue}\right)}{2\sqrt{\left(\text{Red} - \text{Green}\right)^2 - \left(\text{Red} - \text{Blue}\right)\left(\text{Green} - \text{Blue}\right)}} \tag{5}$$

$$S = \frac{\max\left(\text{Red,Green,Blue}\right) - \min\left(\text{Red,Green,Blue}\right)}{\max\left(\text{Red,Green,Blue}\right)} \tag{6}$$

$$V = \max\left(\text{Red,Green,Blue}\right) \tag{7}$$

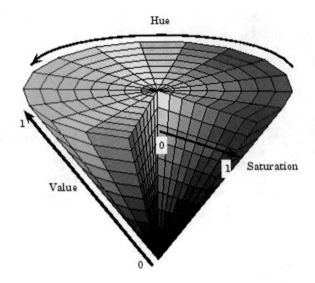

Fig. 6 HSV color model [1]

Using these three equations (5), (6), and (7), convert RGB color space into HSV color space.

2.3 YCbCr Model

This is a luminance-based color model. A nonlinear RGB signal is shown in Fig. 7 which is in an encoded form of it. Although this model is extensively exercised in European television studios, it is also used in image compression-related work. In this model, color is defined in the form of luma which is more sensitive to the human eye (luminance computed from nonlinear RGB). Moreover, the RGB values are formed in a weighted sum of that colors in [4]. The blue component shows C_b and red component illustrates C_r values. These are related to the chroma component and now these are the less sensitive to the human eye. The transformation of luminance and chrominance components is simple and explicit for the separation of these color components which make a perfect YCbCr color space [3].

$$Y = 0.299\text{Red} + 0.587\text{Green} + 0.114\text{Blue}$$
$$C_b = \text{Blue} - \text{Yellow}$$
$$C_r = \text{Red} - \text{Yellow} \tag{8}$$

Fig. 7 YCbCr color model [1]

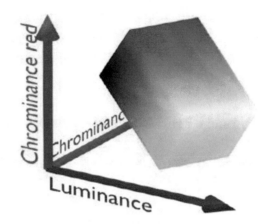

Fig. 8 CMYK model [1]

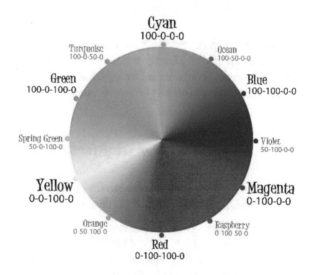

2.4 CMYK (Cyan, Magenta, Yellow, and Black) Model

This is a subtractive color model as illustrated in Fig. 4. It is often used in color printing technology. In this color model, cyan is a shade of blue; generally it is considered as cold blue or gray sky blue, and the magenta color is in the form of additive synthesis; it is a combination of a blended red and blue. In a similar fashion, yellow is also analogous to a painter's yellow, although it is a bit paler. Black color is using a black ink, and it is economical and does not utilize the colored inks. The transformation from RGB in CMYK color space is clearly shown in Eq. (9) (Fig. 8).

$$Cyan = 1 - Red$$
$$Magenta = 1 - Green\}$$
$$Yellow = 1 - Blue \qquad\qquad (9)$$

3   Image Processing and Color Models

Humans are able to perceive thousands of colors but only a couple of dozen gray shades.

3.1   Pseudo-color Image Processing

Terms pseudo-color and false color are both the same in image processing. It is commonly applied to identify the method of color assignment to a single colored image. However, humans can recognize more 100 color shades and intensities in comparison with two dozens of gray shades.

3.2   Gray Level to Color Transformation Method

This color transformation technique into in gray level is very common, but it gives us better enhancement results of pseudo-color processing.

Intensity to Color Transformation
Figure 9 illustrates the functional block diagram of color transformation for pseudo-color processing. In this case, we are using a function of color which is further divided into three functions that are blue, red, and green. These separated color functions of blue, red, and green are fed into the RGB color monitor as the corresponding inputs (red, blue, and green) and the corresponding output for each color is produced.

3.3   Full-Color Image Processing

This section deals with a processing technique which is applicable to full-color image processing. This image processing technique comes in two major classes:

1. Under the first category, we process each and every element of image and after that form a composite processed image. As a result, this image is formed by individually processed elements.

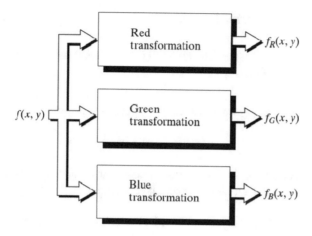

Fig. 9 Block diagram of color transformation for pseudo-color image processing

2. Under the second category, we are able to work with non-monochrome pixel straightway and here every pixel is considered as a vector.

We are assuming C is an arbitrary vector for red, green, and blue color space:

$$W = \begin{bmatrix} c_r & R \\ c_g & G \\ c_b & B \end{bmatrix} \tag{10}$$

Equation (10) illustrates that these elements of W are only red, green, and blue elements of a "color image processing." These color elements are a function of (x, y) coordinates.

$$W = \begin{bmatrix} c_r(a,b) & R(a,b) \\ c_g(a,b) = G(a,b) \\ c_b(a,b) & B(a,b) \end{bmatrix} \tag{11}$$

The image size is $M \times N$ which is vector form.

$$W(x,y), \quad \text{where,} \ a = 0,1,2,...,M-1; \ b = 0,1,2,...,N-1$$

Equation (11) shows a vector and its elements are in the form of variables a and b. The formulation of these color elements is simple and concedes to process a color image with its each element individually using the method of grayscale image processing.

3.4   Color Transformation

The process of color image elements with a single color model is known as color transformation technique.

Formulation
The expression of gray-level transformation technique is given below:

$$g(a,b) = \text{Trn}\big[f(a,b)\big] \tag{12}$$

where we considered $f(a, b)$ is a function of color input image, $g(a, b)$ is the function of transformed color output image, and T is an operator on f over a spatial neighborhood of $f(a, b)$.

Showing the images can choose the pixel values from color space as in three or four groups.

$$X_i = \text{Trn}\big(r_1, r_2, \ldots, r_k\big), \qquad i = 1, 2, \ldots, k \tag{13}$$

where, r_i and $X_i \longrightarrow$ Both are variables for the color components of $f(a, b)$ and $g(a, b)$ at any point (a, b); $k \longrightarrow$ Total color components; Trn $\longrightarrow \{T_1, T_2, \ldots, T_k\}$: "set of transformation"/"color mapping functions" that will work on r_i and will generate X_i.

A picture of a color image of a bowl of strawberries and a coffee cup is illustrated in Fig. 10 and it is also in a color negative form. The shown picture is in a high resolution. In the next row, all color components are in CMYK model, after that it is in RGB, and finally it is in HIS model. Hence, any kind of transformation can be performed on whatever color model.

4   Related Work

Our literature survey which is based on color detection techniques found various approaches. Some of them we consider here and revealed their findings. Many human skin detection techniques have been used earlier on different color spaces. Bernhard Fink et al. (2006) [10] applied skin color and texture detection techniques which were important cues for people to consciously or unconsciously differentiate the various culture-based views about each other. Skin color and texture could be a sign of beauty, race, health, age, wealth, etc. Jones et al. (2002) [11] demonstrated the human face detection distribution using the three Gaussian cluster methods for human skin colors in the YCbCr color space. Chai and Bouzerdoum (2000) [12] have proposed an image classification technique based on Bayesian approach to classify all pixels into categories, namely, skin color and non-skin color. In another

Fig. 10 Various color space components of full-color image. (Original image courtesy of Meddata Interactive)

work we found a different detection approach in which the color spaces are used with separated luminance and chrominance components for detecting human face. Gomez et al. (2002) [14] applied the attribute selection technique which belongs to machine learning community; this approach can select appropriate color space for skin detection. Albiol et al. (2001) [15] have revealed that if any optimized skin detector system is designed for each color model, the outcome of all skin detector systems is the same because of the optimization of the systems. Furthermore, Kolkur et al. (2016) [16] used another method based on segmentation method which considers that the RGB channels or other color space combinations can take all color spaces. It can also allow to differentiate the different color and texture combinations in that particular image.

5 Proposed Methodology

Skin Modeling

The main objective of our work is to detect skin pixels and non-skin pixels based on conditioning rule, for example, if the certain condition, which is given in the proposed algorithm, is satisfied, then choose x or otherwise go for y.

The process flow diagram of the proposed model is presented in Fig. 12. Few factors are required to detect skin color and determine the threshold value. Firstly, segregate skin-colored pixels and non-skin-colored pixels in the source color image or test image, device, or related hardware or camera which would be used to take an image; however, the outputs of the same test image would be different [6]. Furthermore, skin tone varies from one person to another. The result of the object movement is in a blurred image, and the shadow and brightness effects on color of the image. The illumination in the image may get change, so we need to check it.

Color Detection Based on Image Segmentation

Strong human skin color segmentation involves mathematical model to specify the skin color dispensation. Segmentation procedure is quite a tedious task and it has a unique color range (all color combination). "Skin color and non-skin color" segmentation is shown in Fig. 11.

Skin detector involves some basic steps:

A. *Let x be a 2D pixel array of converted image in the RGB channel.*
B. *Collect an arbitrary photographs from the Internet.*
C. *Convert image to a 2D pixel array.*
D. *Extracted image size is 128 × 128 and each pixel has 32 bits.*
E. *Choose a suitable color space.*
F. *Convert image in the same color space.*
G. *Based on threshold, classify each pixel using the skin classifier whether it is a "skin pixel or non-skin" pixel.*

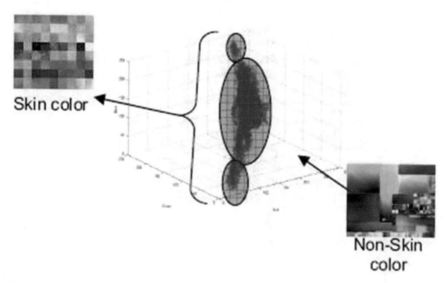

Skin color

Non-Skin color

Fig. 11 Segmentation of skin and non-skin color

H. *Skin classifier learning parameters, i.e., TP ("true positive"), TN ("true negative"), FP ("false positive"), FN ("false negative"), W ("overall measure of classification").*

5.1 Proposed Skin Detection Algorithm

The ranges for a skin pixel in different color spaces used by the proposed algorithm are as follows:

Red > 95 and Green > 40 and Blue > 20 and Red > Green and Red > Blue and I Red-Green I > 15 and A > 15

or

Red > 95 and Green > 40 and Blue > 20 and Red > Green and Red > Blue and I Red-Green I > 15 and Crmin > 135 and
Cb > 85 and Ylumin > 80 and Crmin <= (1.5874*Cb) +20 and
Crmin >= (0.3447*Cb) +76.2068 and
Crmin >= (−4.5652*Cb) +234.5652 and
Crmin <= (−1.15*Cb) +301.78 and
Crmin <= (−2.2868*Cb) +433.85 nothing and
Key < 205 and 0 <= Cyan <= 0.05 and 0.088 < Yellow
<1 and 0 <= Cyan/Yellow < 1 and 0.1 <= Yellow/Magenta < 4.8 or
Key < 205 and 0 <= Cyan <= 0.05 and 0.0909 <
Y < 0.945 and 0.1 <= Yellow/Magenta < 4.67

or

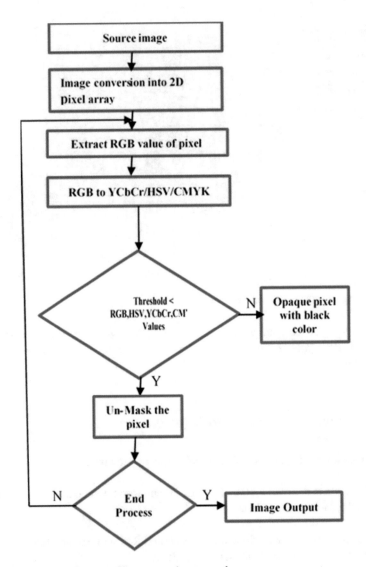

Fig. 12 Process flow diagram of the proposed system color

For cyan level:
0 <= C <= 0.005
R: "Red"; B: "Blue"; G: "Green";
Cr, Cb: "Chrominance components"; Y: "luminance component",
C: "cyan"; M: "magenta"; Y: "yellow"; K: "key" (black).
The flow diagram of the proposed work is illustrated in Fig. 12.

Table 1 Restrictions for RGB, YCbCr, and HSV color spaces

Color space	Scale of components	Restrictions for skin color
RGB	Red, Green, Blue: [0,255]	Red > 95 and Green > 40 and Blue > 20
YCbCr	Ylumin, "Cb, Cr": [0,255]	Ylumin >80 and 77 < Cb < 127 and 133 < Cr < 173
HSV	H: [0,360], S, V: [0,1]	0 < H < 50 and 0.1 < S < 0.68 and 0.35 < V < 1

5.2 Restriction of Different Color Spaces

Restriction of the color components in RGB, YCbCr, and HSV space for human skin detection is shown in Table 1.

6 Results and Discussion

6.1 Comparative Analysis of Different Color Models

The performance evaluation of different color spaces is based on thresholding method. This skin detection method obtains results from randomly chosen data from the Internet. The table below shows the summary of skin detection using various color spaces. Table 2 illustrates all parameters like true positive rate (TPR), false positive rate (FPR), false negative rate (FNR), and true negative rate (TNR) different color model configuration. As we know for better performance, the parameters such as TPR and TNR should be high and FNR and FPR should be low. After analyzing this experimental result, HSV is the best color model for skin detection compared to another color model. Although the total number of pixels of the original image is 108506, 51188 skin color pixels were detected and also 57318 non-skin color pixels were detected.

The best performance achieved in terms of accuracy is 79% in the case of HSV model. Moreover, the comparative analysis in terms of "TPR" (sensitivity), "FPR" (1-specificity), and "FNR," "TNR" (specificity), and W with other novel work is also provided in Table 2. To achieve an exceptional performance, the value of *TPR and TNR* should be high, whereas the value of *FNR and FPR* should also be low.

$$Accuracy = \frac{T_N + T_P}{T_N + F_P + T_P + F_N} = 79\%$$

$$Sensitivity\left(TPR\right) = \frac{T_P}{T_P + F_N} = 87.2\%$$

$$Specificity\left(TNR\right) = \frac{T_N}{T_N + F_P} = 87\%$$

Table 2 Result of performance parameter of skin color detection with different color space

Color space	RGB (%)	HSV (%)	YCbCr (%)	CMYK (%)
TPR	42	86	21	87
FPR	8	33	8	75
FNR	57	13	78	12
TNR	92	67	92	24
Overall measure (W)	51	82	35	75

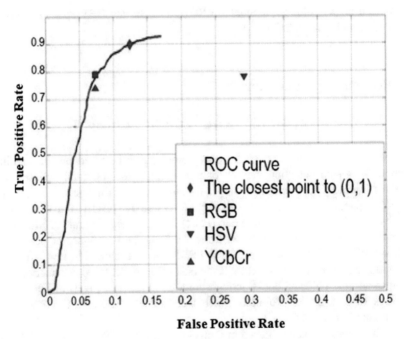

Fig. 13 Graph of receiver operator characteristics (ROC) of different detection algorithms

6.2 ROC Analysis

The receiver operator characteristic (ROC) curve analysis is often used to evaluate the quality of the approach of skin color detection algorithm. Moreover, these selected points are going to be marked as the corresponding parameters which are obtained from the usage of algorithms. These values can be summarized in Table 2. Hence, these algorithms are working in different spaces such as RGB, HSV, and YCbCr. All parameters like test accuracy and weighted average of precision and recall can also be measured by these algorithms. From the graph it is clear its closest points are zero and one. F1 score can be defined as a combined point of optimal blended values of precision and recall. When the score of F1 is 1, it means it is best and in case it is 0, it is considered as a worst case. Figure 13 demonstrates the receiver operator characteristics in terms of true and false positive values. The closest point of different color space algorithms can be seen from the graph.

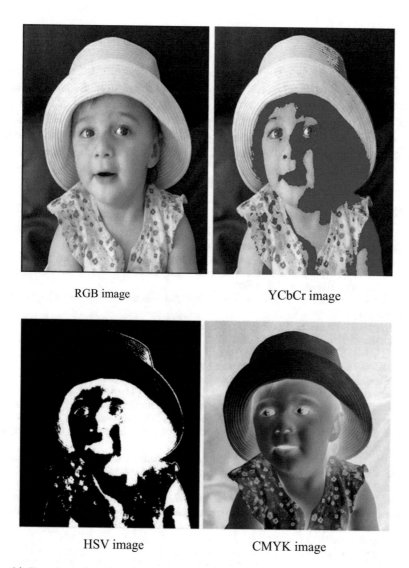

RGB image

YCbCr image

HSV image

CMYK image

Fig. 14 Experimental results on sample image

6.3   Experimental Results of Sample Image

After getting experimental results of the sample image, we can easily analyze the performance of different color models as shown in Fig. 14. All images are the output of various color models.

7   Conclusion

In many applications, skin color detection is playing the key roles, as, for instance, skin disease, gesture tracking, and computer-human interaction face recognition for security purpose. This research focuses on a comparative analysis of human skin detection technique using distinguished color models such as RGB, HSV, YCbCr, and CMYK. According to our results, HSV has better results out of other color models with high TPR and TNR. Finally, the values of FPR and FNR are low.

References

1. Color Basics Homepage., www.colorbasics.com. Last accessed: 22 Nov 2020
2. J.C. Russ, *The Image Processing Handbook*, 3rd edn. (CRC Press, Boca Raton, 1999)
3. B.G. Haskell, A.N. Netravali, *Digital Pictures: Representation, Compression, and Standards* (Perseus Publishing, New York, 1997)
4. M. Petrou, P. Bosdogianni, *Image Processing: The Fundamentals* (John Wiley & Sons, Chichester, 1999)
5. Z. Zhang, H. Gunes, M. Piccardi, Head detection for video surveillance based on categorical hair and skin color models, in 6th IEEE International Conference on Image Processing, pp. 1137–1140 (2009)
6. M. Rasheed, Face detection based on skin color point pixel processing using OpenCV, http://aspilham.blogspot.com/2011/01/face-detection-based-on-skin-colour.html
7. American Cancer Society, *Cancer Facts and Figures 2018* (American Cancer Society, 2018), https://www.cancer.org/content/dam/cancer-org/research/cancer-facts-and-statistics/annual-cancer-facts-and-figures/2018/cancer-facts-and-figures-2018.pdf. Accessed 3 May 2018
8. A.R. Sadri et al., WN-based approach to melanoma diagnosis from dermoscopy images. IET Image Process. **11**(7), 475–482 (2017)
9. C. Barata, M. Ruela, M. Francisco, T. Mendonça, J.S. Marques, Two systems for the detection of melanomas in dermoscopy images using texture and color features. IEEE Syst. J. **8**(3), 965–979 (2014)
10. K.G. Bernhard Fink, P.J. Matts, Visible skin color distribution plays a role in the perception of age, attractiveness, and health in female faces. Evol. Hum. Behav. **27**(6), 433–442 (2006)
11. M.J. Jones, J.M. Rehg, Statistical color models with application to skin detection. Int. J. Comput. Vis. **46**(1), 81–96 (2002)
12. D. Chai, A. Bouzerdoum, A Bayesian approach to skin color classification in YCbCr color space, in Proceedings of the IEEE Region Ten Conference, September 25–27, 2000, Kuala Lumpur, Malaysia (2000)
13. J.C. Terrillon, M.N. Shirazi, H. Fukamachi, S. Akamatsu, Comparative performance of different skin chrominance models and chrominance spaces for the automatic detection of human

faces in color images, in Proceedings of the International Conference on Automatic Face and Gesture Recognition, March 26–30, Grenoble, France (2000)

14. G. Gomez, M. Sanchez, S. Luis Enrique, On selecting an appropriate colour space for skin detection, in Proceedings of the Second Mexican International Conference on Artificial Intelligence: Advances in Artificial Intelligence, April 22–26, Quebec, Canada (2002)

15. A. Albiol, L. Torres, E.J. Delp, Optimum color spaces for skin detection, in Proceedings of the International Conference on Image Processing, October 7–10, Thessaloniki, Greece (2001)

16. S. Kolkur, D. Kalbande, P. Shimpi, C. Bapat, J. Jatakia, Human skin detection using RGB, HSV and YCbCr color models, in *ICCASP/ICMMD-2016. Advances in Intelligent Systems Research*, vol. 137, (Atlantis Press, 2016), pp. 324–332

17. C. Barata, J.S. Marques, J. Rozeira, A system for the detection of pigment network in dermoscopy images using directional filters. IEEE Trans. Biomed. Eng. **59**(10), 2744–2754 (2012)

18. T. Mendonça, P.M. Ferreira, J. Marques, A.R.S. Marcal, J. Rozeira, PH^2 – A dermoscopic image database for research and benchmarking, in 35th International Conference of the IEEE Engineering in Medicine and Biology Society, July 3–7, Osaka, Japan (2013)

19. G. Schaefer, B. Krawczyk, M. Emre Celebi, H. Iyatomi, Melanoma classification using dermoscopy imaging and ensemble learning, in Second IAPR Asian Conference on Pattern Recognition (2013)

20. Z. Waheed, A. Waheed, M. Zafar, F. Riaz, An efficient machine learning approach for the detection of melanoma using dermoscopic images, in International Conference on Communication, Computing and Digital Systems (C-CODE), March 8–9 (2017)

21. Y. Gu, J. Zhou, B. Qian, Melanoma detection based on mahalanobis distance learning and constrained graph regularized nonnegative matrix factorization, in IEEE Winter Conference on Applications of Computer Vision (WACV), March 24–31 (2017)

22. T.Y. Satheesha, D. Satyanarayana, M.N. Giri Prasad, K.D. Dhruve, Melanoma is skin deep: A 3D reconstruction technique for computerized dermoscopic skin lesion classification. IEEE J. Transl. Eng. Health Med. **5**, 4300117 (2017)

23. A. Abi-Dargham, E. van de Giessen, M. Slifstein, et al., Baseline and amphetamine-stimulated dopamine activity are related in drug-naïve schizophrenic subjects. Biol. Psychiatry **65**(12), 1091–1093 (2009)

24. A. Smirnov, *Processing of Multidimensional Signals* (Springer-Verlag, New York, 1999)

25. J. Goutsias, L. Vincent, D. S. Bloomberg (eds.), *Mathematical Morphology and Its Applications to Image and Signal Processing* (Kluwer Academic Publishers, Boston, 2000)

26. S. Marchand-Maillet, Y.M. Sharaiha, *Binary Digital Image Processing: A Discrete Approach* (Academic Press, New York, 2000)

27. S.E. Umbaugh, *Computer Vision and Image Processing: A Practical Approach Using CVIP tools* (Prentice Hall, Upper Saddle River, 1998)

28. K.R. Castleman, *Digital Image Processing*, 2nd edn. (Prentice Hall, Upper Saddle River, 1996)

29. B. Jahne, *Digital Image Processing: Concepts, Algorithms, and Scientific Applications* (Springer-Verlag, New York, 1997)

30. P.M. Mather, *Computer Processing of Remotely Sensed Images: An Introduction* (John Wiley & Sons, New York, 1999)

31. A.H. Mallot, *Computational Vision* (The MIT Press, Cambridge, MA, 2000)

32. S. K. Mitra, G. L. Sicuranza (eds.), *Nonlinear Image Processing* (Academic Press, New York, 2000)

33. S. Edelman, *Representation and Recognition in Vision* (The MIT Press, Cambridge, MA, 1999)

A Study of Improved Methods on Image Inpainting

Ajay Sudhir Bale, S. Saravana Kumar, M. S. Kiran Mohan, and N. Vinay

1 Introduction

Recent studies have shown that inpainting the missing regions in an image is a challenging task [1]. The images are required to generate the visual structures of images and also the textures. In reality, these approaches often produce distorted and blurry textures which are inconsistent with the neighboring areas. The main reason for these problems is the ineffectiveness of the neural networks which is evolutional as the borrowing of information from spatial locations is not possible in all the conditions. When the textures are to be borrowed from the surrounding areas, patch syntheses are the most suitable approaches. Image inpainting can restore the old photographs and can remove the superimposed text like subtitles and date. It can also be used to remove the objects entirely. The given image u has the masked region Ω, filling all the pixels in the value which is taken from Ω^c [2]. Image inpainting is not only restoring and modifying the images but also the understanding of validity of image models. As these models have various inpainting techniques, basically divided into three categories, one category considers the repetitive nature of local information.

In many cases the texture is modeled considering one pixel has brightness values which are independent from the rest of the image [3]. The neighborhood area is nothing but a square patch which is around the pixel and whose size is an important parameter of the used algorithm. The model images are the inputs and the real challenge is to find the difference between various patches to predict the value. This kind of approach is one-pass greedy algorithm. In this algorithm, an order procedure is designed having priorities on edge strength [4]. The representation of image sparse is a new approach to the inpainting problem [5]. The method is to present the images in sparse combination of sets such as wavelets, DCT, etc. After this, the

A. S. Bale · S. S. Kumar · M. S. Kiran Mohan · N. Vinay (✉)
Department of ECE, SoET, CMR University, Bengaluru, India

© Springer Nature Switzerland AG 2022
P. Johri et al. (eds.), *Trends and Advancements of Image Processing and its Applications*, EAI/Springer Innovations in Communication and Computing,
https://doi.org/10.1007/978-3-030-75945-2_15

missing pixels are concluded by updating this representation [6]. In [7], the text was improved by separating the text with that of cartoons.

The recent developments in deep learning and GAN-based approaches made it well suitable for solving the image inpaint problem. The classical image inpainting approaches use either local or nonlocal information to frame the missing regions. If the region is small, the local information helps to generate the missing areas. But if the area of missing region is large, then we can make use of the patches from a nonlocal information, an external database, to generate the missing areas.

In the recent past, features like image levels have been used to overcome the inpainting problem [8]. In order to reconstruct the area which was missing in the image, Patch Match method was used [9–11]. This method gives real texture features only, as it uses the low-level features of the given image and the high-level features cannot be extracted. This method uses differential equations for extracting the features. The repair algorithms which use partial differentiation are cyclic coordinate descent (CCD) [8, 12–14] and the total variation algorithm [15–17]. The damaged image is repaired using diffusion to the information which is local in the image. If the image is large, and the repairing area is large, it gives different effects. The texture information-based repair uses the best block of image from the area boundary which is to be repaired.

A popular technique of K-singular value decomposition (K-SVD) [9] is used for image denoising. The deep learning method of denoising the astronomical images is also popular [18]. The Criminisi algorithm uses the sparse technique wherein the discontinuous edges are also repaired [19]. Image inpainting has a capability of recovering the blocks, removing the unwanted text and objects as shown in Fig. 1.

Fig. 1 Image inpainting capabilities

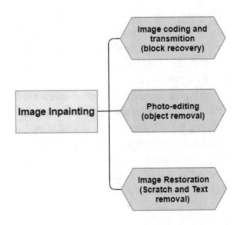

2 Comparative Study

2.1 Image Inpainting

Image inpainting is a process of automatic regaining of images when the images are holes due to confidential and artistically considerations. Mainly there are two types of image inpainting that are PDE [20–23] and exemplar approach. The PDE-based techniques are typically created to connect edges resulting in boundary data discontinuities and also to enhance the lines in the inpainting domain in an appropriate manner.

The work of the inpainting on the image depends on the criteria on which the image has been destructed and the function that caused the destruction to the image [22]. The process of inpainting has been widely used in the removal of undesired objects that are visual in the image. The inpainting strategies can be grouped into two; they are non-exemplar method [24–27] and exemplar method [23, 28, 29]. The non-exemplar approach is based on the pixel intensity continuity that is pixel interpolation; this method is more expensive and less effective when there is a large missing region. So, the exemplar-based method is widely used in recent years, as this method can extract complex structures in the missing area of the image.

The text is generated through exemplar method based on the similarity of the pattern, differentiated between the missing and proper region of the image. In some cases, feature space distance, wavelet domain, eigenspace, and Fourier space are used for similarity measurement [23, 28–30]. In some cases, the sum of squared differences (SSD)-based identical pattern measure is used [31–34]. The non-parametric sampling has been used to copy the data to the missing region from the data region [31]; this method is capable of generating the missing structures that are complex, but the order of the copied texture will be affected. To achieve proper texture, mappings such as fixed pixel number in a window, capacity isophotes, and similarity of pattern have to be defined for the order of texture copy [31, 33, 34].

The interactive global-based optimization method is employed in the painting methods to avoid problems of ordering [35–37]. The objective function in this technology that assesses the similarity of the pattern is specified and optimized using the graph cut approach, EM algorithm, and belief propagation. The global optimization technique results effectively in a variety of images, but in some cases improper images are still outputted. This is due to the limitation of data region sampling and lack of pattern similarity for generating natural textures. The image accuracy can be improved by concentrating on these two factors [23]. In some studies, to overcome this issue, the orientation and scale of textures have been used [38]. Changing the brightness of the sampling textures can be done to obtain the effective samplings. In case of limited data region sampling and the locality of spatial texture patterns is treated as an implicit constraint that is commonly met the real scenes in the case of lack of pattern similarity, that is implemented with the energy minimization framework [23].

The PatchMatch approach is a type of image inpainting technology, designed for reconstructing the missing part of the image [9–11]. The technique of PatchMatch is capable of providing details about real texture, but it just uses a provided context's low-level characteristics and lacks the capacity to predict high-level background characteristics. The sparse representation method is advanced to improve the standard of image repair that helps to get the improved, unblurred images. In recent years many methods have been introduced to improve the quality [9] of the image using advanced techniques like direct learning technique developed based on K-singular value decomposition (K-SVD) [9], combination of sparse representation and Criminisi algorithm for decreasing the amount of sparse calculation [19], collaborative deep learning (CDL) technique to remove the noise from astronomical image [18], collaborative of K-SVD and fuzzy C-means clustering for image recovering [39], super-resolution reconstruction-based sparse representation for image denoising [40], and online dictionary learning method where the model is trained for sparse encoding [41] that is later updated with the optimized dataset [8, 42].

In order to generate semantic information more efficiently from the image, the integration of neural networks and deep learning has been suggested in recent years. The context encoding [8, 40] has been used for encoding damaged information that is pre-detected; this method has shown the effective results outputting the better repaired images, but the technique also lacks a few repairing information and shows visible repair traces [43–47]. In the case of higher-resolution image repairing, adding of convolution layer will make the receptive field higher and integration of global and local discrimination network will increase the repair consistency [48]; although it has shown the effective results for high-resolution images, it still has some limitation [49–54]. By integrating the generative network and SE-DResNet (squeeze-and-excitation network deep residual learning) [55] module, the clarity of the repaired image can be improved and made realistic [8].

2.2 Sparse Representation

The filling order consists of an input image, wherein a target region is selected by the user [56, 57]. This target region has to be removed and again filled. This can be easily stated by a user. The images are grown from the removed portion to the inside. At each stage, the priority on pixel is computed. If the patch is lying on the boundary, then some of the pixels are said to in the target region. Now these patches are understood to be a signal in the region of source and the pixels are represented by the lost components. These patches are recovered by sparse representation. After each recovery, the boundaries are updated.

EM algorithm is used for inpainting of images based on linear sparse representations [58]. EM algorithm is represented by x ¼ Fa [58, 59], in which the image x has very few atoms in the dictionary. This makes the imposing of penalty on the reconstructed coefficients. Thus, inpainting is a problem which occurs due to missing data and this EM is a very general means of framework in such situations. The missed data are replaced by the estimated coefficients. Then this process is repeated until it gets converged. The Bayesian EM algorithm [60] is one in which a sequence is produced which estimates two steps: namely, E-step and M-step. In E-step, the expected values are found from the observed complete data. In M-step, denoising [61] operation is carried out based on the penalty function.

2.3 Markov Random Field Modeling and Multiscale Graph Cuts

The global methods (in which inpainting is defined as the global optimization function) use Markov random field (MRF) [62–65]. In [62], a smart algorithm is used to optimize the objective function, which removes the unwanted labels by visiting them in a proper order. In [66] the MRF method is used to improve the inpainting process. The processes involve improving and increasing the speed of search in general. A novel method of optimization is used to make the MRF inpainting suited to a large number of labels. To deduce the formal properties [67] of machine vision, the MRF model is most popular. The model has rich filters, and a trained database is used for training the generic images with the help of contrastive divergence. Such a model is most useful in any interference method which requires in prior the spatial images.

The algorithm in [68] gives good results in very less time. In image processing type MRFs, a pixel is simply assumed to take a gray level of their surrounding pixels. In this kind of processing, the images produced are very smooth and clear. In [69], a non-local-based MRF model is used to train the model. By using this model, nonlocal filters learn the local structure in natural images.

The algorithm which is based on MRF by using the stochastic model [70] reconstructs the texture by taking only one pixel at the first attempt [71–74]. Then all the individual pixels are queried by with the use of a metric called as the sum of squared differences (SSD). The method in which the area is marked by following the same angle as that of isophote line follows the edges of the images proposed in [75].

In MRF, the process begins with the user asking for selecting the area in the image which has to be reconstructed. This is accomplished by checking all the points on the boundary of the image. These checked points are stored in a file and then connected as a line by joining all the dots. The pixels which are inside the polygon as specified by the user are treated as the area which has to be reconstructed [70]. The next part is reconstructing each pixel by using stochastic method. MRF is used to find the very next value of the sequence which can be used in bioinformatics [76, 77], speech recognition [75], and energy consumption modeling [78].

The exemplar-based approach is used for the energy boost problem [79]. This energy parameter has a term which makes sure that the reconstructed images that are placed at the boundaries of the inpainted domain have a good continuation. This method is used to get global optimum images with the help of the multiscale graph cut (MGC). The energy problem can be minimized using a vector representation which compares the patches in low resolution and reduces the information loss.

In [73], a priori technique is introduced to improve the search of similar patches using multiscale approach. The missed areas are approximated by using a level from course to fine levels [80]. The input picture with coarse version can be inpainted much easily when compared with that of the full resolution.

2.4 Neural Networks and GAN

Contextual attention [81] is a feedforward fully convolution neural network which can inpaint images with multiple holes with different size and different locations. The method utilizes the surrounding information of the image to generate proper image structure in the missing region.

Globally and locally consistent image completion [1, 82] is a method that uses two content discriminators. Global discriminator helps to specify the filled part that is coherent with the whole image, and local discriminator checks the completeness of a small area of the completed region. The algorithm generates portions which are not present anywhere in the image, so it is well suitable for generating the portions of the faces. The algorithm is not efficient for inpainting images with heavily structured objects. The context encoders [83] are convolution neural networks used to generate the missing parts of an image conditioned to its surrounding regions. But it can detect the semantics of the image, thereby helping to inpaint effectively.

Multi-Scale Neural Patch Synthesis [84] uses an approach a multi-scale neural patch synthesis approach to fill the missing regions from an input image. The algorithm removes problems like blurry and improper boundaries from learning-based approaches. In the case of high-resolution images, the algorithm creates sharper regions. The disadvantage of the algorithm is that it will create discontinuity whenever the image has a complicated structure and the cost is another constraint.

The generative adversarial network (GAN) was first used by Goodfellow [85]. The model has two parts, namely, the generator and the discriminator. The generator is used to generate the images and uses this image to map the image with the help of random noise. The discriminator is used to differentiate the true and false samples from the collection of inputs. Figure 2 shows the GAN model.

The image of certain kind is identified by using the real image features. The gap between the real and the generated images is minimized by GAN.

When the surrounding area of the image is used to repair by spreading the texture information, only the missing images in the small area are repaired. The PatchMatch technique uses the method to find the best matching block to minimize the trace of repair effect. Although this method provides the texture information, it cannot predict the structure of the object. GAN has a high performance in the high-resolution image production, but training the GAN network in order to generate high-definition images is very difficult.

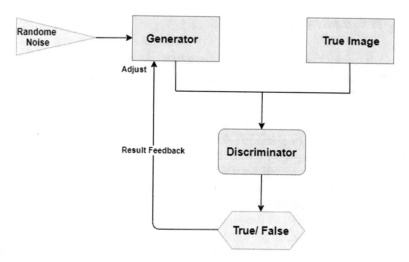

Fig. 2 GAN model

Recent training methods like progressive GANs [44] provide good results to train the network and are highly capable of producing high-resolution images. The context encoder [40] improves the repairing of the images. The discriminator trains the network and determines the generated image is true or false. At this time, the discriminator is updated. The mean squared error (MSE) along with loss function is used to repair 64*64-pixel image [8] which is the middle of 128*128-pixel image and repairs the network. This will overcome the problem of blurring when MSE is lost.

High-Resolution Image Synthesis and Semantic Manipulation with Conditional GANs [86] is also suitable to synthesize realistic images using conditional GANS, suitable for high-resolution images [87].

Deep generative model [88] uses GAN model for predicting appropriate content for the missing region from the training data, thereby creating sharper edges which in turn makes the region more realistic. The algorithm is well suited for images with simple structures like face, not efficient for the images with complex structures. The prediction is purely based on the generative model and training.

Boosted GAN [89] is a GAN-based approach which incorporated semantic information of the image for the inpainting process to reduce the ambiguity of the prediction done by the traditional approaches. To discover the semantic and attribute information, there are two auxiliary networks used in inpainting. The discriminator ensures the consistency of semantic attributes with respect to that of original image.

2.5 *Texture Synthesis*

Texture is a frequently used term in the process of computer vision, having high practical value. In the field of research, texture synthesis is used most widely. The texture synthesis techniques are mostly categorized to example-based and procedural-based text synthesis. The synthesis of procedural texture is directly based on a curved plane by simulating physical objects [90] to overcome the disturbance due to texture mapping [91–93]. In example-based method [94–99], large regions of texture images are created by the process of sampling texture in smaller regions depending on the self-similarity of the texture. In procedural process, system parameters must be readjusted every time depending on the image; this is overcome by example-based method [90].

In example-based method there are main two types; they are pixel-based method and patch-based method. In pixel-based methods, the image's texture can be represented as the interrelationship in the image between the pixels [90]. The overview of this method is to evaluate and reproduce interactions between each pixel [90]. Pixel-based approaches have the following functions: local conditional, likelihood estimation of each pixel's density, and the generation of a new pixel-by-pixel image. That depends on searching the best fit. Every pixel is identified by looking at the image that is inputted, depending on the nearby pixels that are almost similar, which have been pre-synthesized in the image that is outputted [90]. In recent years patch-based methods have improved drastically, for example, spatial coherence overcomes blurring artifact [100], neighborhood matching technique based parallel controllable algorithm for texture synthesis [97].

In patch-based method, each time a patch of pixel is copied, the patch size is defined depending on the texture feature. Patch-based method will identify the complete patch's best match, by searching for the best match by analyzing the overlap among the new and synthesized patches. Hence, the latest image is created patch by patch. The random patch-pasting-based texture synthesis algorithm provides faster generation of the textures patch by patch [101].

The real-time patch-based sampling synthesis process helps to prevent mismatching of the functionality through patch limits and also offers seamless transition between the adjacent patches of the texture [102]. Liang's approach with Efros's algorithm decreases the mismatching features around the patch limits by cutting the minimum error boundary [103]. In deformation and feature matching-based texture synthesis, deformation and curve features are employed to increase the efficiency of identifying the matching patch [104]. New optimization-based approaches have been developed by inter-mating both the pixel-based and patch-based approaches [105], introduced in recent years, combining the advantages of both pixel and patch methods [106].

The Directional Empirical Mode Decomposition (DEMD) [107–110] had advanced in recent years in terms of texture classification and segmentation that takes picture path into consideration in decomposition and helps to extract three feature values for individual pixel. Hence, the DEMD is self-adaptive and entirely data driven. The method of iteration is adopted for extracting individual component. So, DEMD is self-adaptive [90] locally and provides a distractive benefit in retrieving data for visual perception. DEMD-based algorithm for texture synthesis [90] performs effectively in synthesizing the higher-quality images based on the feature values; this algorithm can also be used for the classification and segmentation of the textures effectively [108, 109].

The advance development of image processing technology in recent years had made the integration of images with data [111]. One such technique is QR code; it has been generated as an accessible and reliable 2D stable bar code and has been popularly used for integrating the URL (uniform resource locator). The pointless QR-code binary pattern harms the artistic value of the printed images that are overcome by cooling process. In recent years, as a supplementary of QR code, the natural images have been embedded with data [111]. The embedding of the data depends on the frequent channel and the color component specifically with some numerical techniques advanced as steganography [112, 113] or water markings. The methods of data embedding on printing media have the same properties like that of steganography and have the ruggedness against attacks. The use of re-coating and random coating will enhance the consistency of the synthesized [114–116] texture image from the initial LBP (local binary pattern) painting [111].

2.6 Feature Distribution

The feature set can be based on gray-level difference method [117]. The popular methods are DIFFX and DIFFY, wherein the histograms of gray level have differences in pixels of neighbors in horizontal and vertical directions [118]. In the case of DIFF2, the differences are found in vertical and horizontal directions, but in DIFF4 the differences are in all the four principal directions. Thus, these differences provide the measures for texture rotation. The GLC matrix method is used for defining the neighboring pixels of any image [119]. The two considered elements of this matrix have frequencies which have two image pixels which are kept apart from each other by a pixel distance. Both of these pixels have separate intensities. But this method uses a lot of memory space and the time required for calculation is more.

The Voronoi tessellation is the property used to take out the texture tokens of a given image [120]. This name is given as the distribution of local spatial is present in Voronoi polygon shapes. These texture tokens can be the points with high gradient or a close boundary in the image. Any token which has a neighbor is defined by the Voronoi polygon. Most of the features of a token are present in the properties of these neighborhoods. For the application of geometrical methodologies to the gray-level images, the tokens are extracted. Each cell of Voronoi is extracted and the similar characteristics of tokens are formed in one group to build a region of uniform texture [120].

3   Conclusion

The sparse representation gives high consistency in texture and noise for the holes which are large in shape. The unwanted pixels are removed from the digital graph and are called as less greedy algorithm.

The EM approach also uses sparse representation to recover the missing samples in the computation. EM approach is considered best for zooming and also for the tasks of disocclusion.

The MRF modeling is much faster compared to gray-level approach and also it consumes less memory. It is said to be the best approach for the removal of text or in the complete removal of objects. The MRF approach can be used when the images are having problems like demosaicing; also it has the problem of super-resolution.

MGC solves the energy minimization problem very efficiently by the introduction of patches at low resolution. This method can minimize or eliminate the probability of inexactness. It also improves the offset map accuracy. The MGC approach is good when the main and dominant structures are to be retrieved as the luminance and the orientation is less numerous.

The capability of GAN to frame new images helps the process of inpainting with more realistic images. The problem of the traditional approach is the ambiguity of the prediction. The [89] method incorporates the semantic information of the input image to frame the missing regions which results in better image compared to that of the real image. [84, 88, 91] are well suitable for the images with simple structures and not efficient with the complex object structures. [84, 86] will create more sharp regions for the high-resolution images.

The exemplar method of repairing the images has greater performance when compared to non-exemplar methods. With the use of deep learning, the efficiency is drastically increased in recent times.

The pixel-based method and patch-based method for the texture synthesis have evolved in recent years with their own advances, but with the advance of optimization-based approaches, the pixel and patch method has seen higher efficiencies. The introduction of re-coating and random coating has increased the quality of image texture synthesis which could even replace the QR code technology.

References

1. J. Yu, Generative image inpainting with contextual attention, in *The IEEE Conference on Computer Vision and Pattern Recognition (CVPR)*, (2018), pp. 5505–5514
2. A. Bugeau, M. Bertalmío, V. Caselles, G. Sapiro, A comprehensive framework for image inpainting. IEEE Trans. Image Process. **19**, 2634–2645 (2010)
3. A. Efros, T. Leung, Texture synthesis by non-parametric sampling. Proc. IEEE Int. Conf. Comput. Vision **2**, 1033–1038 (1999)

4. A. Criminisi, P. Pérez, K. Toyama, Region filling and object removal by exemplar-based inpainting. IEEE Trans. Image Process. **13**(9), 1200–1212 (2004)
5. Z. Xu, J. Sun, Image inpainting by patch propagation using patch sparsity. IEEE Trans. Image Process. **19**(5), 1153–1165 (2008)
6. Z. Li, H. He, H. Tai, Z. Yin, F. Chen, Color-direction patch-sparsity-based image inpainting using multidirection features. IEEE Trans. Image Process. **24**(3), 1138–1152 (2015). https://doi.org/10.1109/TIP.2014.2383322
7. O.G. Guleryuz, Nonlinear approximation based image recovery using adaptive sparse reconstructions. IEEE Int. Conf. Image Process. (2003)
8. Y. Chen, L. Liu, J. Tao, et al., The improved image inpainting algorithm via encoder and similarity constraint. Vis. Comput. (2020). https://doi.org/10.1007/s00371-020-01932-3
9. M. Aharon, M. Elad, A. Bruckstein, K-SVD: An algorithm for designing overcomplete dictionaries for sparse representation. IEEE Trans. Signal Process. **54**(11), 4311–4322 (2006)
10. J. Zhang, Y. Wu, W. Feng, J. Wang, Spatially attentive visual tracking using multi-model adaptive response fusion. IEEE Access **7**, 83873–83887 (2019)
11. Y. Chen, W. Xu, J. Zuo, K. Yang, The fire recognition algorithm using dynamic feature fusion and IV-SVM classifier. Cluster Comput. **22**, 7665–7675 (2019)
12. J. Zhang, S. Zhong, T. Wang, H. Chao, J. Wang, Blockchain-based systems and applications: A survey. J. Internet Technol. **21**(1), 1–14 (2020)
13. J. Wang, J. Qin, X. Xiang, Y. Tan, N. Pan, CAPTCHA recognition based on deep convolutional neural network. Math. Biosci. Eng. **16**(5), 5851–5861 (2019)
14. D. Altantawy, A. Saleh, S. Kishk, Texture-guided depth upsampling using Bregman split: A clustering graph-based approach. Vis. Comput. **36**, 333–359 (2020)
15. C. Yang, H. Feng, Z. Xu, Q. Li, Y. Chen, Correction of overexposure utilizing haze removal model and image fusion technique. Vis. Comput. **35**, 695–705 (2019)
16. Y. Liu, J. Pan, Z. Su, K. Tang, Robust dense correspondence using deep convolutional features. Vis. Comput. **36**, 827–841 (2020)
17. B. Yin, X. We, J. Wang, N. Xiong, K. Gu, An industrial dynamic skyline based similarity joins for multi-dimensional big data applications. IEEE Trans. Ind. Inform. **16**(4), 2520–2532 (2020)
18. Beckouche, S., Starck, J., Fadili, J.: Astronomical Image Denoising Using Dictionary Learning. (2013). arXiv arXiv:1304.3573
19. Hu, G., Ling, X.: Criminisi-based sparse representation for image inpainting. In: Proceedings of IEEE International Conference on Multimedia Big Data, Laguna Hills, CA, USA, 19–21 April 2017, pp. 389–393 (2017)
20. M. Bertalmio, G. Saporo, V. Caselles, C. Ballester, Image inpainting, in *SIGGRAPH*, (2000)
21. R. Tibshirani, Regression shrinkage and selection via the Lasso. J. R. Stat. Soc. Series B Stat. Methodol. **58**(1), 267–288 (1997)
22. O. Elharrouss, N. Almaadeed, S. Al-Maadeed, et al., Image inpainting: A review. Neural. Process. Lett. **51**, 2007–2028 (2020). https://doi.org/10.1007/s11063-019-10163-0
23. N. Kawai, T. Sato, N. Yokoya, Image inpainting considering brightness change and spatial locality of textures and its evaluation. Lect. Notes Comput. Sci, 271–282 (2009). https://doi.org/10.1007/978-3-540-92957-4_24
24. A. Levin, A. Zomet, Y. Weiss, Learning how to inpaint from global image statistics. Proc. ICCV **1**, 305–312 (2003)
25. C. Ballester, V. Caselles, J. Verdera, M. Bertalmio, G. Sapiro, A variational model for filling-in gray level and color images, in *Proceedings ICCV*, (2001), pp. 10–16
26. E. Villéger, G. Aubert, L. Blanc-Féraud, Image disocclusion using a probabilistic gradient orientation, in *Proceedings ICPR*, vol. 2, (2004), pp. 52–55
27. S. Esedoglu, J. Shen, Digital ipainting based on the Mumford-shah-euler image model. Eur. J. Appl. Math. **13**, 353–370 (2003)
28. A.N. Hirani, T. Totsuka, Combining frequency and spatial domain information for fast interactive image noise removal, in *Proceedings SIGGRAPH*, vol. 1996, (1996), pp. 269–276

29. T. Amano, Image interpolation by high dimensional projection based on subspace method, in *Proceedings ICPR*, vol. 4, (2004), pp. 665–668

30. S.D. Rane, J. Remus, G. Sapiro, Wavelet-domain reconstruction of lost blocks in wireless image transmission and packet-switched, in *Proceedings ICIP*, vol. 1, (2002), pp. 309–312

31. R. Bornard, E. Lecan, L. Laborelli, J. Chenot, Missing data correction in still images and image sequences, in *Proceedings of ACM International Conference on Multimedia*, (2002), pp. 355–361

32. A.A. Efros, T.K. Leung, Texture synthesis by non-parametric sampling, in *Proceedings ICCV*, (1999), pp. 1033–1038

33. B. Li, Y. Qi, X. Shen, An image inpainting method, in *Proceedings of IEEE International Conference on Computer Aided Design and Computer Graphics*, (2005), pp. 531–536

34. A. Criminisi, P. Pérez, K. Toyama, Region filling and object removal by exemplar-based image inpainting. Trans. Image Process. **13**(9), 1200–1212 (2004)

35. Y. Wexler, E. Shechtman, M. Irani, Space-time completion of video. Trans. PAMI **29**(3), 463–476 (2007)

36. C. Alléne, N. Paragios, Image renaissance using discrete optimization, in *Proceedings ICPR*, (2006), pp. 631–634

37. N. Komodakis, G. Tziritas, Image completion using global optimization, in *Proceedings CVPR*, (2006), pp. 442–452

38. I. Drori, D. Cohen-Or, H. Yeshurun, Fragment-based image completion, in *Proceedings SIGGRAPH*, vol. 2003, (2003), pp. 303–312

39. S. Darabi, E. Shechtman, C. Barnes, D. Goldman, P. Sen, Image melding: Combining inconsistent images using patch-based synthesis. ACM Trans. Graph. **31**(4), 1–10 (2012)

40. Pathak, D., Krahenbuhl, P., Donahue, J., Darrell, T., Efros, A.: Context encoders: feature learning by inpainting. In: Proceedings of IEEE Conference on Computer Vision and Pattern Recognition, Las Vegas, NV, USA, 27–30 June 2016, pp. 2536–2544 (2016)

41. Y. Luo, J. Qin, X. Xiang, Y. Tan, Q. Liu, L. Xiang, Coverless real-time image information hiding based on image block matching and dense convolutional network. J. Real-Time Image Process. **17**(1), 125–135 (2020)

42. T. Naderahmadian, S. Beheshti, M. Ali, Correlation based online dictionary learning algorithm. IEEE Trans. Signal Process. **64**(3), 592–602 (2015)

43. W. Li, H. Xu, H. Li, Y. Yang, P. Sharma, J. Wang, S. Singh, Complexity and algorithms for superposed data uploading problem in networks with smart devices. IEEE Internet Things J. (2019). https://doi.org/10.1109/JIOT.2019.2949352

44. H. Yang, Z. Zhang, Depth image upsampling based on guided filter with low gradient minimization. Vis. Comput. **36**, 1411–1422 (2020)

45. N. Liao, Y. Song, X. Huang, J. Wang, Detection of probe flow anomalies using information entropy. J. Intell. Fuzzy Syst. (2020). https://doi.org/10.3233/IFS-191448

46. G. Sheng, X. Tang, K. Xie, J. Xiong, Hydraulic fracturing microseismic first arrival picking method based on non-subsampled shearlet transform and higher-order-statistics. J. Seism. Explor. **28**(6), 593–618 (2019)

47. F. Yu, L. Liu, S. Qian, L. Li, Y. Huang, C. Shi, S. Cai, X. Wu, S. Du, Q. Wan, Chaos-based application of a novel multistable 5D memristive hyperchaotic system with coexisting multiple attractors. Complexity **2020**, 8034196 (2020)

48. S. Iizuka, E. Simo-Serra, H. Ishikawa, Globally and locally consistent image completion. ACM Trans. Graph. **36**(4), 107:1–107:14 (2017)

49. Y. Chen, J. Wang, S. Liu, X. Chen, J. Xiong, J. Xie, K. Yang, Multiscale fast correlation filtering tracking algorithm based on a feature fusion model. Concurr. Comput. Pract. Exp. (2019). https://doi.org/10.1002/cpe.5533

50. Z. Liao, J. Peng, Y. Chen, J. Zhang, J. Wang, A fast Q-learning based data storage optimization for low latency in data center networks. IEEE Access **8**, 90630–90639 (2020)

51. F. Yu, L. Liu, H. Shen, Z. Zhang, Y. Huang, C. Shi, S. Cai, X. Wu, S. Du, Q. Wan, Dynamic analysis, circuit design and synchronization of a novel 6D memristive four-wing hyperchaotic system with multiple coexisting attractors. Complexity **2020**, 5904607 (2020)
52. E. Mikaeli, A. Aghagolzadeh, M. Azghani, Single-image superresolution via patch-based and group-based local smoothness modeling. Vis. Comput. (2019). https://doi.org/10.1007/s00371-019-01756-w
53. N. Pan, J. Qin, Y. Tan, X. Xiang, G. Hou, A video coverless information hiding algorithm based on semantic segmentation. EURASIP J. Image Video Process. (2020). https://doi.org/10.1186/s13640-020-00512-8
54. Nie, G., Cheng, M., Liu, Y., Liang, Z., Fan, D., Liu, Y., Wang, Y.: Multi-level context ultra-aggregation for stereo matching. In: Proceedings of IEEE Conference on Computer Vision and Pattern Recognition, Long Beach, CA, USA, 16–20 June 2019, pp. 3283–3291 (2019)
55. Hu, J., Shen, L., Sun, G.: Squeeze-and-excitation networks. In: Proceedings of IEEE Conference on Computer Vision and Pattern Recognition, Salt Lake City, UT, USA, 18–22 June 2018, pp. 7132–7141 (2018)
56. J. Lee, D. Lee, R. Park, Robust exemplar-based inpainting algorithm using region segmentation. IEEE Trans. Consum. Electron. **58**(2), 553–561 (2012). https://doi.org/10.1109/TCE.2012.6227460
57. B. Shen et al., Image inpainting via sparse representation, in *International Conference on Acoustics, Speech, and Signal Processing*, (2008)
58. A. Dempster, N. Laird, D. Rubin, Maximum likelihood from incomplete data via the EM algorithm. J. R. Stat. Soc. B **39**, 1–38 (1977)
59. M.J. Fadili, J.L. Starck, F. Murtagh, Inpainting and zooming using sparse representations. Comp. J. **52**(1) (2009)
60. J.M. Fadili, J.L. Starck, EM algorithm for sparse representation-based image inpainting, in *IEEE International Conference on Image Processing*, (2005). https://doi.org/10.1109/icip.2005.1529991
61. C. Wu, On the convergence properties of the em algorithm. Ann. Stat. **11**, 95–103 (1983)
62. N. Komodakis, G. Tziritas, Image completion using efficient belief propagation via priority scheduling and dynamic pruning. IEEE Trans. Image Process. **16**(11), 2649–2661 (2007)
63. T. Huang, S. Chen, J. Liu, X. Tang, Image inpainting by global structure and texture propagation, in *Proceedings of ACM International Conference on Multimedia*, (2007), pp. 517–520
64. Y. Yang, Y. Zhu, Q. Peng, Image completion using structural priority belief propagation, in *Proceedings of ACM International Conference on Multimedia*, (2009), pp. 717–720
65. T. Rûzíc, A. Pîzurica, W. Philips, Markov random field based image inpainting with context-aware label selection, in *ICIP '12*, (2012), pp. 1733–1736
66. T. Rûzíc, A. Pizurica, Context aware patch based image inpainting using Markov random field modelling. IEEE Trans. Image Process. (2014). https://doi.org/10.1109/TIP.2014.2372479
67. S. Roth, M.J. Black, Fields of experts: A framework for learning image priors, in *IEEE Computer Society Conference on Computer Vision and Pattern Recognition (CVPR'05)*, (2015)
68. Y.M. Ohkubo, K. Tanaka, Digital image inpainting based on Markov random field, in *International Conference on Computational Intelligence for Modelling, Control and Automation*, (Web Technologies). https://doi.org/10.1109/cimca.2005.1631558
69. J. Sun, M.F. Tappen, Learning non-local range Markov random field for image restoration, in *CVPR*, (2011)
70. A. Gellert, R. Brad, Image inpainting with Markov chains. SIViP **14**, 1335–1343 (2020). https://doi.org/10.1007/s11760-020-01675-7
71. A. Efros, T. Leung, Texture synthesis by non-parametric sampling, in *7th IEEE International Conference on Computer Vision*, (Corfu, Greece, 1999), pp. 1033–1038
72. O. Elharrouss, N. Almaadeed, S. Al-Maadeed, Y. Akbari, Image inpainting: A review. Neural Process. Lett., 1–22 (2019)
73. C. Guillemot, O. Le Meur, Image inpainting: Overview and recent advances. IEEE Signal Process. Mag. **31**(1), 127–144 (2014)

74. P. Patel, A. Prajapati, S. Mishra, Review of different inpainting algorithms. Int. J. Comput. Appl. **59**(18), 30–34 (2012)
75. M. Bertalmio, G. Sapiro, V. Caselles, C. Ballester, Image inpainting, in *27th Annual Conference on Computer Graphics and Interactive Techniques*, (New Orleans, 2000), pp. 417–424
76. V. Jääskinen, V. Parkkinen, L. Cheng, J. Corander, Bayesian clustering of DNA sequences using Markov chains and a stochastic partition model. Stat. Appl. Genet. Mol. Biol. **13**(1), 105–121 (2014)
77. A. Mushtaq, C.-H. Lee, An integrated approach to feature compensation combining parti-cle filters and hidden Markov model for robust speech recognition, in *IEEE International Conference on Acoustics, Speech, and Signal Processing*, (Kyoto, 2012), pp. 4757–4760
78. A. Gellert, A. Florea, U. Fiore, F. Palmieri, P. Zanetti, A study on forecasting electricity pro-duction and consumption in smart cities and factories. Int. J. Inf. Manag. **49**, 546–556 (2019)
79. Y. Liu, V. Caselles, Exemplar-based image Inpainting using multiscale graph cuts. IEEE Trans. Image Process. **22**(5), 1699–1711 (2013)
80. I. Drori, D. Cohen-Or, H. Yeshurun, Fragment-based image completion. ACM Trans. Graph. **22**(2003), 303–312 (2003)
81. J. Yu, Z. Lin, J. Yang, X. Shen, X. Lu, T.S. Huang, Generative image inpainting with contex-tual attention. CVPR (2018)
82. S. Iizuka, E. Simo-Serra, H. Ishikawa, Globally and locally consistent image completion. ACM Trans. Graph. (TOG) **36**(4), 107 (2017)
83. D. Pathak, P. Krahenbuhl, J. Donahue, T. Darrell, A.A. Efros, Context encoders: Feature learning by inpainting. CVPR (2016)
84. C. Yang, X. Lu, Z. Lin, E. Shechtman, O. Wang, H. Li, High resolution image inpainting using multi-scale neural patch synthesis. CVPR (2017)
85. Goodfellow, I., Pouget-Abadie, J., Mirza, M., Xu, B., Warde-Farley, D., Ozair, S., Courville, A., Bengio, Y.: Generative adversarial nets. In: Proceedings of Annual Conference on Neural Information Processing System, Montreal, Quebec, Canada, 7–12 December 2015, pp. 5672–2680 (2015)
86. T. Wang, M. Liu, J. Zhu, A. Tao, J. Kautz, B. Catanzaro, High-Resolution image synthesis and semantic manipulation with conditional GANs, in *IEEE/CVF Conference on Computer Vision and Pattern Recognition*, (2018), pp. 8798–8807
87. S. Ding, J. Zheng, Z. Liu, Y. Zheng, Y. Chen, X. Xu, J. Lu, J. Xie, High-resolution dermoscopy image synthesis with conditional generative adversarial networks. Biomed. Signal Process. Control **64**, 102224, ISSN 1746-8094 (2021). https://doi.org/10.1016/j.bspc.2020.102224
88. R.A. Yeh, C. Chen, T.Y. Lim, A.G. Schwing, M. Hasegawa-Johnson, M.N. Do, Semantic image inpainting with deep generative models. CVPR (2017)
89. A. Li, J. Qi, R. Zhang, R. Kotagiri, Boosted GAN with semantically interpretable informa-tion for image inpainting, in *International Joint Conference on Neural Networks (IJCNN)*, (Budapest, Hungary, 2009), p. 1
90. Y. Zhang, Z. Sun, W. Li, Texture synthesis based on direction empirical mode decompo-sition. Comput. Graph. **32**(2), 175–186, ISSN 0097-8493 (2008). https://doi.org/10.1016/j.cag.2008.01.001
91. J. Dorsey, A. Edelman, J. Legakis, et al., Modeling and rending of weathered stone, in *Proceedings of ACM SIGGRAPH*, (ACM Press, Los Angeles, 1999), pp. 225–234
92. A. Witkin, M. Kass, Reaction-diffusion textures, in *Proceedings of ACM SIGGRAPH*, (ACM Press, Los Angeles, 1991), pp. 299–308
93. S. Worley, A cellular texture basis function, in *Proceedings of ACM SIGGRAPH*, (ACM Press, New Orleans, 1996), pp. 291–294
94. L.Y. Wei, M. Levoy, Fast texture synthesis using tree-structured vector quantization, in *Proceedings of ACM SIGGRAPH*, (ACM Press, Los Angeles, 2000), pp. 479–488
95. A.A. Efros, T.K. Leung, Texture synthesis by non-parametric sampling, in *International Conference on Computer Vision*, (ACM Press, Greece, 1999), pp. 1033–1038

96. P. Harrison, A non-hierarchical procedure for resynthesis of complex textures, in *WSCG 2001 Conference Proceedings*, (2001), pp. 190–197
97. S. Lefebvre, H. Hoppe, Parallel controllable texture synthesis, in *Proceedings of the SIGGRAPH*, (Los Angeles, 2005), pp. 777–786
98. Y.X. Liu, Y. Tsin, W.C. Lin, The promise and perils of near-regular texture. Int. J. Comput. Vis. **62**(1–2), 145–159 (2005)
99. P. Zhang, S.L. Peng, Structure-based texture synthesis. J. Comp. Aided Design Comp. Graph. China **16**(3), 290–296 (2004)
100. M. Ashikhmin, Synthesizing natural textures, in *2001 ACM Symposium on Interactive 3D Graphics*, (ACM Press, Los Angeles, 2001), pp. 217–226
101. Y. Xu, B. Guo, H.Y. Shum, *Fast and Memory Efficient Texture Synthesis. Technical Report: MSR-TR-2000-32* (Microsoft Research, Beijing, 2000)
102. L. Liang, C. Liu, Y. Xu, et al., Real-time texture synthesis using patch-based sampling. ACM Trans. Graph. **20**(3), 127–150 (2001)
103. A.A. Efors, W.T. Freeman, Image quilting for texture synthesis and transfer, in *Proceedings of ACM SIGGRAPH*, (ACM Press, Los Angeles, 2001), pp. 341–347
104. Q. Wu, Y. Yu, Feature matching and deformation for texture synthesis. ACM Trans. Graph. (TOG) **23**(3), 364–367 (2004)
105. H. Shen et al., Missing information reconstruction of remote sensing data: A technical review. IEEE Geosci. Remote Sens. Magaz. **3**(3), 61–85 (2015). https://doi.org/10.1109/MGRS.2015.2441912
106. V. Kwatra, I. Essa, A. Bobick, et al., Texture optimization for example-based synthesis, in *Proceedings of the SIGGRAPH*, (Los Angeles, 2005), pp. 795–802
107. Z.X. Liu, H.J. Wang, S.L. Peng, Texture segmentation using directional empirical mode decomposition, in *2004 International Conference on Image Processing (ICIP)*, (2004), pp. 279–282
108. Z.X. Liu, S.L. Peng, Directional empirical mode decomposition and its application to texture segment. Sci. China Series E Inf. Sci. **35**(2), 113–123 (2005)
109. Z.X. Liu, H.J. Wang, S.L. Peng, Texture classification through directional empirical mode decomposition, in *Proceedings of the 17th International Conference on Pattern Recognition (ICPR'04)*, (2004), pp. 803–806
110. M.K. Jha, S.D. Roy, B. Lall, DEMD-based video coding for textured videos in an H.264/MPEG framework. Pattern Recogn. Lett. **51**, 30–36 (2015). https://doi.org/10.1016/j.patrec.2014.08.010
111. H. Otori, S. Kuriyama, Data-embeddable texture synthesis, in *Smart Graphics. SG 2007*, Lecture Notes in Computer Science, ed. by A. Butz, B. Fisher, A. Krüger, P. Olivier, S. Owada, vol. 4569, (Springer, Berlin, Heidelberg, 2007). https://doi.org/10.1007/978-3-540-73214-3_13
112. N. Provos, P. Honeyman, Hide and seek: An introduction to steganography. IEEE Sec. Priv. **1**(3), 32–44 (2003)
113. Fujii, Y., Nakano, K., Echizen, K., Yosiura, Y.: Digital Watermarking Scheme for Binary Image (in Japanese) Japan Patent 2004-289783
114. M. Ashikhmin, Synthesizing natural textures, in *Symposium on Interactive 3D Graphics*, (2001), pp. 217–226
115. L.-Y. Wei, M. Levoy, Fast texture synthesis using tree-structured vector quantization, in *Proceedings of SIGGRAPH 2000*, (2000), pp. 479–488
116. X. Tong, J. Zhang, L. Lui, X. Wang, B. Guo, H. Shum, Synthesis of bidirectional texture functions on arbitrary surfaces, in *ACM SIGGRAPH 2002*, (2002), pp. 665–672
117. T. Ojala et al., A Comparative study of texture measures with classification based on feature distributions. Pattern Recogn. **29**(1), 51–59 (1996)
118. J. Rombaut, A. Pizurica, W. Philips, Passive error concealment for wavelet-coded I-frames with an inhomogeneous gauss–Markov random field model. IEEE Trans. Image Process. **18**(4), 783–796 (2009). https://doi.org/10.1109/TIP.2008.2011388

119. H. Sarah, Peckinpaugh. Improved method for computing Gray-level cooccurrence matrix based texture measures. Graph. Models Image Process. **52**, 574 (1991)
120. M. Tuceryan, A.K. Jain, Texture analysis, in *The Handbook of Pattern Recognition and Computer Vision*, (World Scientific Publishing Co, 1998), pp. 207–248

Index

© Springer Nature Switzerland AG 2022
P. Johri et al. (eds.), *Trends and Advancements of Image Processing and its
Applications*, EAI/Springer Innovations in Communication and Computing,
https://doi.org/10.1007/978-3-030-75945-2

Printed in the United States
by Baker & Taylor Publisher Services